Lucrèce Funmilayo FALOLOU

Occupancy dynamics and urban heat islands in Porto-Novo

Lucrèce Funmilayo FALOLOU

Occupancy dynamics and urban heat islands in Porto-Novo

ScienciaScripts

Imprint

Any brand names and product names mentioned in this book are subject to trademark, brand or patent protection and are trademarks or registered trademarks of their respective holders. The use of brand names, product names, common names, trade names, product descriptions etc. even without a particular marking in this work is in no way to be construed to mean that such names may be regarded as unrestricted in respect of trademark and brand protection legislation and could thus be used by anyone.

Cover image: www.ingimage.com

This book is a translation from the original published under ISBN 978-620-6-69823-4.

Publisher:
Sciencia Scripts
is a trademark of
Dodo Books Indian Ocean Ltd. and OmniScriptum S.R.L publishing group

120 High Road, East Finchley, London, N2 9ED, United Kingdom
Str. Armeneasca 28/1, office 1, Chisinau MD-2012, Republic of Moldova, Europe
Printed at: see last page
ISBN: 978-620-8-27737-6

Contents

DEDICATION

I dedicate this book to :

♦♦♦ To my very dear mother

An inexhaustible source of tenderness, peace and sac... ice. Your prayers, your bënëdiction and your presence by my side have always been my source of strength to face the various obstacles. No matter what I say, I won't be able to thank you as I should. May God bless you.

♦♦♦ To my dearest father

Who instilled in me a sense of a job well done and to whom I owe this training. May this work express my gratitude and affection. God bless you.

♦ My sweet husband and my lovely daughter Simisola Trinity

No tribute could match the love they have always showered on me. No words could express my gratitude and love. May God protect you and give you health and long life.

Resume

Population growth in urban areas is not without consequences for natural resources. Poor land-use practices (construction of buildings, road and socio-community infrastructure, etc.), including deforestation and agricultural expansion, are deteriorating Benin's natural resources and creating a number of environmental problems, including the phenomenon of urban heat islands (UHI). The general objective of this research is to analyse the influence of (urban) green spaces and building density on the number and intensity of heat islands in the city of Porto-Novo and its surroundings. The methodological approach used is based, on the one hand, on the processing of satellite images and the Geographic Information System (GIS), for the analysis of the dynamics of land use and population evolution and, on the other hand, on the Cellular Automata (CA) Markov model, for the prediction of land use. Digitisation of buildings of all categories from high-resolution Google Earth images and estimation of carbon storage in the aerial biomass of the Jardin des Plantes et de la Nature (JPN) were used to characterise the influence of building density and green spaces on the variation in UHI. The impact of UHI on the population and the environment was determined using night-time surface temperature data from MODIS thermal satellite images obtained via the LPDAAC site from the Unbutu Linux terminal. An investigation was carried out among 870 people to gather information on their perception of heat. This investigation was supported by two series of diurnal temperature measurements at the urban and neighbourhood levels. The results show that areas of natural vegetation are disappearing at the expense of agricultural areas and buildings. Indeed, the remnants of sacred forests have declined sharply, from 0.23% in 1972 to 0.16% in 2012. By 2032, these remnants will occupy just 0.03% of the total surface area, which means that they will tend to disappear over the years. This is not the case for other land use units, in particular settlements and crops. The dynamics of land use and building density are a function of demographic change and population density. The carbon stock in the aerial biomass of the botanical garden conservatory is 463.9 t C/ha (tonne of Carbon per hectare) and that of site 2 (detente) is 337.4 t C/ha, giving a total of 801.3 t C/ha. The JPN alone has the capacity to absorb 801.206 t C/ha. Thus, the more trees there are, the greater the quantity of carbon stored, which easily leads to the formation of cooler zones in the urban environment and a reduction in the intensity of UHI, thanks in particular to the reduction in greenhouse gases. Analysis of the distribution of UHIs shows that there is a correlation between the amount of heat measured and the observed building density. Spatial variations in temperature are small, and the thermal divide is reduced in the study area. However, the UHIs in the study area can be classified into three categories: - Low UHI with a temperature difference of 1°C from the surrounding area, medium UHI with a temperature difference of 2°C from the surrounding area and high UHI with a temperature difference of 3°C to 6°C from the surrounding area. High UHI night-time temperatures range from 24°C to 32°C and are found on building roofs, dark impermeable surfaces and asphalt, which ranks them as the hottest surfaces. Similarly, the diurnal temperatures of high UHIs vary from 30°C to 33.5°C and are concentrated in non-vegetated areas, including town centres, traffic lights, shopping areas, traffic areas and roadsides. These increases in night-time and daytime temperatures are linked to urbanisation, and can have not only environmental impacts but also significant health impacts, such as heat stress, heat stroke, fainting spells, etc. Rational management of space is therefore essential, in order to find measures to reduce the phenomenon of UHI.

Key words : Porto-Novo, urban heat islands, building density, green spaces.

General introduction

Cities are often presented as the places where most greenhouse gas emissions and energy consumption are concentrated (Haentjens, 2008). According to the UN, the urban population now outnumbers the rural population and is expected to reach almost five billion people by 2030 (Economic and Social Council, 2007). This growth in the urban population is accompanied by galloping urbanisation, particularly in the countries of the South, which are having to cope with urban growth that is both very sustained and often uncontrolled (Morgane *et al.*, 2012). The extension of urban areas, the increase in the urban population, human activities and the changes they cause to the morphology of a particular area can have devastating consequences for natural environments and the quality of life of citizens (Parmentier, 2010). In fact, changes in surface area and land use (buildings, roads) and changes in the lugosity of surfaces are all parameters that modify the energy balance on a local scale and therefore contribute to generating a particular climate in the city (Oke, 1978). The spatio-temporal variability of the urban climate is the result of the very strong hëtërogënëitë of the urbanised space, with its horizontal and vertical surfaces modifying the physical characteristics of the lower layers of the atmosphere (temperatures, wind, precipitation). These climatic effects, essentially thermal, result from a modification of the energy balance observed within this type of space (Escourrou, 1981; Cantat, 2004).

Thus, one of the observed phenomena generated by human activities is the increase in temperature within an urban environment, compared with the surrounding rural or natural environment. This phenomenon, known as urban heat island, urban thermal island or urban thermodynamic island, is present in most cities (Oke, 1982). The presence of urban heat islands multiplies the episodes of oppressive heat that deteriorate the quality of life in the city. Otherwise, the urban heat island (UHI) effect is a little-known physical climatic phenomenon compared with other manifestations of the same kind, such as the greenhouse effect responsible for climate change. However, on an urban scale, it is just as important, especially as the greenhouse effect reinforces the heat island effect as a driver of climate change, but also on a smaller scale. The heat island effect is generated by the city as a result of its morphology, its materials, its natural, climatic and meteorological conditions, its activities, and so on. But, conversely, it influences the city's climate (temperatures, precipitation), the levels and distribution of pollutants, the comfort of city dwellers and the natural features of cities.

The UHI effect is therefore an urban factor that needs to be taken into account in the design and management of the city. However, it has to be said that the various urban policies are still a long way from taking real account of this phenomenon, which requires - and will require even more in the future if nothing is done today - a reasoned adaptation of the city. At present, the various planning and urban development documents (SDAC, PAO, SCoT, etc.) are still relatively immature on the subject, particularly when compared with equivalent foreign documents. The city of Porto-Novo and its surrounding areas (Adjara, Avrankou, Akpro-Misserete, Agudgud and Seme-Podji), which have undergone remarkable urban development in recent decades, are no exception.

This research focuses on the relationship between urban heat islands (UHI), building density and green spaces in and around the city of Porto-Novo, the political capital of Benin. The aim of this research is to analyse the influence of (urban) green spaces and building density on the evolution of heat islands in the Porto-Novo region. The present thesis therefore reports on the results of the investigations. It is structured in two main parts, subdivided into chapters. The first part presents the theoretical and methodological frameworks. The second part presents the results and recommendations.

PART I

THEORETICAL FRAMEWORK AND METHODOLOGICAL APPROACH

CHAPTER I: THEORETICAL AND CONCEPTUAL FRAMEWORK OF THE RESEARCH

This chapter sets out the research problem, the research questions, the hypotheses, the objectives and the state of knowledge on the subject. In addition, the concepts used and the conceptual framework y are clarified.

1.1 Theoretical framework

It deals with the problem, the hypotheses and objectives, the state of knowledge and the clarification of concepts.

1.1.1. Problem

The urban growth of recent years has resulted in the sprawl of urban areas (Claude, 2012). Today, the city is attracting more and more people. They bring together people and activities. In order to respond to this concentration and to the population's demand for housing, shopping areas, cultural and leisure facilities, the urban space is gradually nibbling away at its peripheral areas (Poulain. 2008, Service de I'Amenagement, de 1'Urbanisme et de 1'Environnement). This is why Amfield (2003) asserts that the densification of buildings in urban areas and, in their periphery, (new residential, commercial and industrial developments) are at the origin of changes in the local climate.

Thus, on the one hand, the urbanisation of space leads to the mineralisation of surfaces, and on the other, the raised structures of buildings modify the local wind regime (increasing the roughness of surfaces, creating canyons). The covering of surfaces by materials, most of which are impermeable and reflect little solar radiation, increases surface temperatures and encourages the formation of heat islands (Amfield, 2003).

In the same vein, Jean-Luc *et al* (2012) argue that the city evolves within a 'natural' environment with which it constantly interacts. The climate is therefore an integral part of this environment, and the built environment has been designed to be as adapted as possible to local climatic conditions.

In addition to being influenced by the climate, cities are also known to influence it. In this way, they modify climatic parameters locally (Morgane *et al.*, 2012). These changes can be observed either by comparison with neighbouring rural areas, or by comparison with their own situation in the past (less urbanised and/or less dense). The city thus induces, within its territory, an increase in temperatures (Landsberg, 1979; Escourrou, 1991) - hence the concept of Urban Heat Island (UHI), - a decrease in wind speed (Sacre, 1983), -- a change in rainfall (Shepherd *et al.*, 2002; Jauregui and Romales, 1996), etc. These changes have consequences for the consumption of energy and the environment. These changes have consequences for the energy consumption of buildings and the efficiency of natural air conditioning (Santamouris *et al.*, 2004), atmospheric pollution (Sarrat *et al.*, 2006; Vautard, 2010), outdoor comfort (Steemers, 2006), health (Buechley *et al.*, 1972) and fauna and flora (Sukopp, 2004).

The term "urban heat island" refers to the difference in temperature observed between urban areas and the surrounding rural areas (Voogt, 2002). Observations have shown that temperatures in urban centres can be up to 12°C higher than in neighbouring regions (Figure 1) (Voogt, 2002).

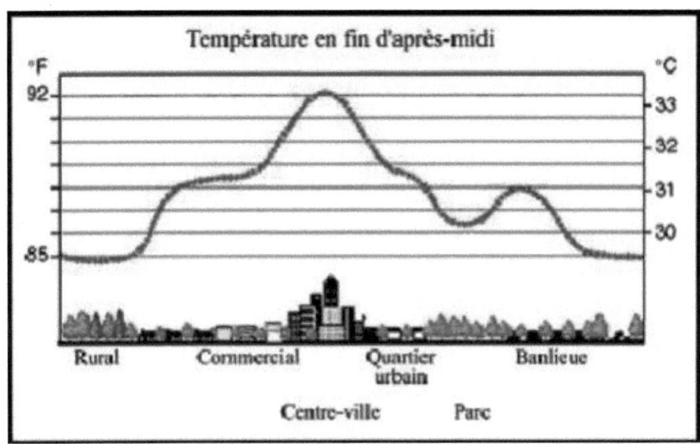

Figure 1 Diagram of the urban heat island
Source: Lawrence Berkeley National Laboratory, 2000.

This phenomenon is in most cases accentuated by urban growth, which is a potential modifier of the climatic conditions of a given region, through the extension of residential buildings, the significant increase in the number of cars, the lack of vegetation, the sealing of spaces, etc. (Givoni, 1989).

Similarly, the intensity of heat islands changes on a daily and seasonal basis as a function of different meteorological and anthropogenic parameters; this is particularly threatening for the urban population (Oke, 1987; Pigeon *et al*, 2008).

In fact, cities such as New York, Chicago, Vancouver, Hong Kong and Geneva, which are constantly growing, have already been suffering for some years from this climatic plienomene, the repercussions of which on city climates are becoming increasingly apparent (Achour, 2006).

The cities of Bënin are part of this trend, particularly the three cities with special status: Cotonou, Parakou and Porto-Novo. This, moreover, is what led Vignon in 2001 to assert that the cities of Bdnin present a common caracleristic in their pënrëlre or, they constitute a conglomërat of bdton, stone and anything other than 1 tree.

This is why many cities have introduced measures to combat urban heat islands. Some cities, such as Paris and Chicago, realised this following heatwaves that proved highly fatal (Besancenot, 2007).

Bëninese cities must also react to these changing climatic realities, in particular by introducing measures to combat urban heat islands and creating urban cool zones. Such initiatives protect the population by increasing its capacity to adapt to such phenomena (Mdlissa, 2009).

The present study focuses particularly on the "Dynamics of occupation and urban heat islands in the city of Porto-Novo and its surroundings (Adjara, Avrankou, Akpro-Missdretd, Aguegue and Sëmë Podji)". In fact, over the last few decades, the city of Porto-Novo, the political and administrative capital, and its surrounding areas, namely: Adjara, Avrankou, Akpro-Missërëtë, Aguegue and Sëmë Podji have undergone remarkable urban development. The built-up area has expanded in a sprawling manner at the expense of natural areas and landscaping. According to data from the results of the RGPH-4 (2013), the city of Porto-Novo is an urban fabric that is home to around 264,320 inhabitants with a growth rate of 2.6% and a density of 5,300 hbts/km2 (RGPH4). Similarly, the communes of Adjara, Avrankou, Akpro-

Missërëtë, Aguegue and Sëmë Podji are home to 97,424 inhabitants with a growth rate of 4.34%, 128.050 inhabitants with a growth rate of 4.42%, 127,249 inhabitants with a growth rate of 4.68%, 44,562 inhabitants with a growth rate of 4.66% and 222,701 inhabitants with a growth rate of 6.09% (RGPH4, 2013).

In addition, the department of I'Oudmd is the third most populated department with a population of 1,100,404 inhabitants, or 11.0% of the total population of the Bdnin (RGPH4, 2013). It is also one of the most urbanised departments, with 44.3% of the population living in cities (RGPH4, 2013).

This spatial growth of these communes is achieved by the settlement of new districts pёпрьёпдиез, as a result of immigration, population movement from the city centre to the përiphërie and housing development (N'Bessa, 1997). This expansion is reflected in an increase in the urban area. Thus, "the countryside is becoming urbanised around Porto-Novo, with houses made of dëfmitive materials in the middle of oil palm plantations, fields of mai's and manioc, as in the north of Ouando on the road to Sakëtë-Pobë ..." (N'Bessa, 1997). (N'Bessa, 1997).

Urban sprawl thus appears to be one of the main consequences of urban density and social intensity (Poulain, 2008).

In the case of the present ëstudy, urban density is nothing more than the density of the built-up area or the density of land occupation. According to МоиНтё *et al.* (2005) the built density (BD) is the ratio between the floor area coefficient (CES); i.e. the ratio between the total floor area of the buildings and the surface area of the block (or surface area of the study area) on which they are трккёз multiplied by the average number of storeys.

*Building footprint * Average height*
Surface area of plot

Built density is not concerned with uses but with the very nature of the subject "the built environment". Built density is an indicator based on what already exists and, in this sense, reflects a lost reality. Thus, an individual visiting a neighbourhood is able to visually appreciate the height and footprint of the buildings, and therefore to have an approximate idea of its built density. In the same way, density, while based on precise indicators, makes it possible to use repёгез to analyse various situations and meet dёvelopment objectives.

However, what we propose to deal with only involves the architectural or gëomëtric dimension of density.

Furthermore, neglected studies on urban climate have rёyёlё that the occurrence of the urban heat island depends mainly on urban density (Givoni, 1989) and that the heat stored in buildings contributes to the rise in urban temperature (Oke, 1988). These variations in tempёrtures are likely to increase the frequency of extreme events such as heat waves, according to the Intergovernmental Panel on Climate Change (IPCC, 2007). In the light of recent events, such as the 2003 heatwave in Europe, cities are proving to be poorly adapted to such hot conditions. So the fundamental question that can be asked here is how to characterise the *"Relationship between the urban heat island, building density and green spaces in the city of Porto-Novo and surrounding area"*. This main question is broken down into the following sub-questions:

- What are parameters explain the variation of 1 етрёгаШге8 in the city of
Porto-Novo and the surrounding area?
- How have UHIs changed between 2001 and 2015?
- How do UHIs impact the environment and people?
- What role do green spaces play in the variation of UHI in the study environment?

Hypotheses have been formulated to answer these questions.

1.1.1.1.   Research hypotheses

The hypoyhёse дёпёгак underlying this research is ibmiulated as follows: building density and green spaces in the city of porto-Novo and its surroundings have an influence on 1 instensity of urban heat islands. This hypothesis is defined as follows:

- 1 land use in the municipality is essentially changing gradually
of Porto-Novo and the surrounding area;
- urban green spaces encourage and contribute to the formation of cooler areas in the commune of porto-Novo and its surroundings, where building density is undergoing an essentially gradual evolution;
- UHIs affect the balance of the urban climate and the health of populations;

1.1.2.2.   Research objectives

The objectives of this study are as follows:

J General objective

The objective дёпёгаl is to study 1 influence of (urban) green spaces and the density of the built plan on the number and intensity of heat islands in the city of Porto-Novo and its surroundings.

J Specific objectives

Spёciflquement il s'agit de :

- Analysing the dynamics of land use in the commune of Porto-Novo and the surrounding area;
- characterising building density and the role and influence of green spaces on UHI ;
- Describing UHIs, their impact on the environment and the population, and proposing sustainable solutions for their regulation in the study environment.

To achieve these objectives, the methodological approach was subdivided into documentary research, fieldwork, data processing and results analysis.

1.1.3. What we know

The state of knowledge revolves around the four sub-themes formulated on the basis of the specific objectives. It addresses issues relating to UHIs in cities, their impact on the environment and strategies for regulating variations in heat as a result of urban planning options.

1.1.3.1.   Urban growth and heat islands

The development of human activities is increasing the greenhouse effect, resulting in a rise in global surface temperature and the risk of major climate change on the planet (Jean-Marc, 2004). According to this author, global warming is none other than the result of human development over many years, to the detriment of the well-being of the environment (...).

M.G.O.P. Obasi (2006), confirms that urban growth is a modifier of the climatic conditions specific to a given region. In fact, he explains that the extension of residential areas, the juxtaposition of apartment blocks, the absence or scarcity of green spaces, the significant increase in the number of cars, exhaust gases, street lighting, heating systems and the sealing off of spaces give rise to a whole series of factors that significantly modify the climate of cities.

Moreover, Givoni (1989) has shown that studies carried out on urban climate have revealed that the appearance of the urban heat island depends mainly on urban density.

In this respect, Oke (1988) confirms that heat is stored in buildings, which contributes to the rise in temperature in the urban environment.

1.1.3.2.   Urban morphology and thermal regulation

Urban morphology, through the three-dimensional shapes, orientation and spacing of buildings in a city, plays a role in the formation of urban heat islands (USEPA, 2008). It is for this reason that Coutts *et al* (2008) assert that tall buildings and narrow streets can hinder the proper ventilation of urban centres, as they create canyons where the heat generated by solar radiation and human activities accumulates and remains trapped. In addition, Oke (1988) states that urban morphology can also influence car traffic, thereby encouraging the input of heat and air pollution from this mode of transport, the motor vehicle.

For Colombert (2008), the **urban heat island (UHI)** phenomenon, also known as the 'urban heat island', refers to a metropolitan area where the temperature is significantly higher than in the surrounding rural areas.According to Colombert (2008), this observed variation in temperature is the result of natural factors and anthropogenic factors (caused by humans), with a predominance of factors specific to built-up areas, such as the absence of trees and vegetation, the presence of large non-renewable surfaces that absorb and store solar energy, and the emission of multiple energy discharges. For this reason, Gigubre (2009) asserts that during the day, non-reflective surfaces such as asphalt store heat from solar radiation and release it at night. In this sense, Colombert (2008), attests that the intensity of urban heat islands varies on a daily and seasonal basis. He explains that the intensity of the heat island is at its highest during the night. It will also be higher following a sunny day with relatively low wind speed.

To illustrate this point made by Colombert (2008), Cavayas *et al* (2008) state that in Quebec, the northern climate greatly reduces the number of heat islands in winter. In btb, the many hours of sunshine mean that July is very often the most favourable period for the formation of heat islands.

1.1.3.3.   Urban heat islands and well-being

Local warming due to urbanisation has many negative impacts on the environment and health (Philippe A. *et al.*, 2011). According to studies carried out in Canada in 2018, Trottier notes that the heat island phenomenon increases the frequency, duration and intensity of **extreme heat waves.** He points out that since the 1980s, natural resources have been adversely affected by the high temperatures recorded. Bourque and Simonet (2007) concur with this view, stating that forecasts show that average temperatures will continue to rise over the coming decades.

In the same vein, the IPCC (2001) states that the heat waves caused by this urban heat island phenomenon entail major public health risks, since they affect the morbidity and mortality rate of the exposed population by creating **heat stress** in individuals, which could prove fatal.

Gigubre (2009), states that extreme heat can cause a range of discomforts and ailments, or exacerbate a chronic *condition* to the point of causing death. He was able to demonstrate that the people most vulnerable to heat are those suffering from chronic illnesses, socially isolated populations, very young children, workers, people from low socioeconomic backgrounds, high-level outdoor sportsmen and women, people suffering from mental disorders and the elderly.

Furthermore, Akbari *et al* (2001) report that urban heat islands contribute to the formation of "smog", hence the worsening of **atmospheric pollution.** Indeed, for the latter and according to Environment Canada (2011), smog, by dëfmition is "a toxic тёlanдe of gases and particles that 1 can often observe in 1 air in the form of seelie haze" and consists mainly of two pollutants: troposphëric ozone (O3) (1 ozone measured at ground level) and fine particles. They also confirm that ozone needs heat to form.

In the same vein, Salomon and Aubert (2003) noted that heat influences the quality of indoor air, as it encourages the multiplication of dust mites, moulds and baeleries. Based on these findings, the Conseil Rëgional de l'Environnement de Montréal (2007), claims that atmospheric pollution is responsible for 1540 premature deaths per year in Montreal. What's more, the increase in acute respiratory symptoms, emergency room visits and cases of bronchitis is also due to the rise in concentrations of pollutants in the air, according to Bouchard and Smargiassi (2007). In addition, ASSSM (2009) agrees that, in Montreal, 6,000 cases of acute bronchitis in children and 114,000 people with asthma symptoms are recorded every day.

In France, Besancenot (2005) reports that in 2003, during the summer months, the country was hit hard by a devastating heatwave. There were 15,000 deaths, with the most vulnerable groups being older people, especially women.

Labrecque and Vergriete (2008, part 3: 3) have also shown that urban heat islands can also affect aquatic ecosystems by raising the temperature of rainwater that comes into contact with their surfaces. This phënomëne then contributes to raising the tempëratures of the nearby streams, rivers, ëtangs and lakes into which they run off.

In this sense, 1 EPA (2011), confirms that an increase in the tempërature of the aquatic environment can then negatively affect the mëtabolism and reproduction of many espëces.

Following the same line of thought, Giguere (2009) attests that urban heat islands lead to an **increase in** drinking **water consumption** and contribute to an **increase in energy consumption**. In fact, he explains that this results in an increase in ënergëtic demand induced by air conditioning/ventilation.

Cavayas and Baudouin (2008) state that urban forest cover has been steadily declining in Quebec since the 1960s, and could even be threatened with extinction within 20 years in the Montreal metropolitan community. In fact, they show that the progressive densification of cities and the development of urban infrastructures in recent decades are the main causes. This is why Bolund and Hunhammar (1999), confirm that this loss of vëgëtation implies a loss of freshness in the urban environment.

1.1.3.4. Urban forestry and thermal regulation

According to Akbari *et al* (2001), English *et al* (2007) and Cavayas and Baudouin (2008), vegetation plays an essential role in protecting against heat through the process of evapotranspiration and shading of soil and buildings. In fact, during the natural process of evapotranspiration of water vapour, the ambient air cools down by giving up some of its heat to allow evaporation to take place. For these authors, vëgëtation also contributes to good rainwater management and better air quality in cities.

Again for these authors, a change in land use from a vëgëtalisëe area to a built-up area gënëre done a "thermal dëgradation".

Cavayas and Baudouin (2008) conducted a comparative study covering the period 1984-2005, which clearly illustrates the progression of mineralisation at the expense of natural areas in the Montreal metropolitan region, particularly in the west of the island of Montreal, on the North Shore and the South Shore. In particular, this study reveals a correlation between elevated temperatures in heavily minëralised areas and the low vegetation cover index in certain boroughs or municipalities.

Similarly, another earlier study using the RAMS model (Dubreuil *et al.*, 2001) showed the increase in temperatura and decrease in relative humidity consëcutifs to dëforestation.

These authors agree with Baudouin *et al* (2007) in showing that areas that have been heavily minëralisëed are much warmer than areas that have retained significant vëgëtation.

Finally, in Bënt, Nathanael (2004) in his тёто!re, measured the urban pressure of the wetlands in the two valleys of the Zounvi and the Воиё (Porto-Novo) gradually won by 1 urban sprawl, with regard to the issue of urban management. His work is based on the notions of urban pressure and transformations of wet ёcosystёmes. Indeed, after making l'ёlяl of the locations and ёcosystёmes of these valleys, he identified the indicators of urban pressure and then ёtablish a description of each indicator.

The various approaches presented here provide, in a general context, the different explanations that have been given and the approaches to solutions that have been proposed in response to the issues raised by our study.

1.1.4. Clarification of some concepts

This ёtude involves a number of concepts that are important to clarify. These include:

Urban morphology: the three-dimensional shape of a group of buildings and the spaces they create (Nikolopoulou, 2004).

This is ёalso the study of urban forms (www.techno-science.net, consulted on 22/11/2015). These two dёfmitions correspond well to our framework of study.

Albedo: According to Nikolopoulou, (2004), l'albedo is dёfmit as ё!anl the fraction of incident solar radiation reflected by a surface.

It is also, a fraction of solar radiation reAёlё by a surface or object, often expressedёe in percentage (**Bessemoulinet, 2000**). In the context of this research, it is the quantity of solar radiation under the effect of gases, buildings, and the low coverage of green space in the research environment that is repelledёe.

Thermal capacity: Amount of heat stored when its temperature increases by 1°C. It is expressed in Wh/m3 °C and is obtained by multiplying the mass by the specific heat of the material. The greater the specific heat, the greater the amount of heat required to raise the temperature of a material (Outils solaires, 2009).

The thermal capacity of a body is a quantity that quantifies the ability of a body to absorb or release energy by heat exchange during a transformation in which its temperature varies (http://www.techno- science.net/onglet=glossary&definition=1307 , consulted on 29/10/2019). These two definitions are used in our study.

Convection: Dёplacement of heat within a fluid by the movement of all its molecules (Outils solaires, 2009).

Heat transport by displacement of matter. Thermal convection, turbulent convection, currents, convective motions. Convection is only possible in fluids where a system of hot currents invades the cold fluid and vice versa (Michard, 1966). These two definitions are used in our study.

Adaptation: gradual growth of the organism's response to repeated exposure to a stimulus, y including all the actions that make it better able to survive in such an environment (Nikolopoulou, 2004).

According to the Larousse French dictionary (www.larousse.fr, consulted on 19/01/2016), adaptation is 1 action of adapting or adapting to something: Adaptation to circumstances.

These two definitions correspond well to the concerns of our study.

Urban forestry: according to Costello (1993), urban forestry is dёfmit as ё!an! "the amёnagement of trees in urban areas".

For Nilson and Randrup (1996), urban forestry is dёflnit as ё!anl: "the planting, management, planning and design of trees and forest stands of amenity value, зкиёз within or adjacent to urban areas.".

In the present ёtude, urban forestry takes into account not only the amёnagement of vёgёtaux but also the management of trees in urban areas.

Espacevert : Il dësignun place where nature is amënagëepour I'agrement et I^panouissement de l'espëce humaine. Les espaces verts embellissent les cites et sont interdits a toutes les ticlivites pouvant les dëgrader (www.francetop.net, consulte le 19/01/2016).

It is also any public space, ркпlё of flowers, trees or grassland intended to preserve or envelop natural and human resources and provide rest, dëtente, oxygënation and recreation (Tente, 2008).

These two definitions correspond well to the concerns of our study.

Jardin public : According to the dictionnaire de 1'urbanisme et de l'amënagement (2005) it is "a green urban space, enclosed, predominantly planted, protected from traffic, free of trees, adjoining a public facility and managed as such". This definition corresponds well to the concerns of our study.

It is also an enclosed or delimited indoor or outdoor space where dëcorative plants (flowers, ornamental trees, etc.) are grown (www.wikrpedia.org, consuhd on 26/05/2016).

Smog: This is "a toxic mixture of gases and particles that can often be observed in the air in the form of a dry haze" and consists mainly of two pollutants: tr	oposphëric ozone (O3) (ozone measured at ground level) and fine particles (Akbari *et al*, 2001) and Environment Canada (2011).

It is also a cloud of atmosphëric pollution made up of particles from combustion (coal-fired power stations, exhaust gases) and troposphëric ozone (http://m.futura-sciences.com/magazines/ , consuhd on 21/10/2015).

These two definitions correspond well to the concerns of our study.

Climate change: According to a report by the Intergovernmental Panel on Climate Change (IPCC, 2001), climate change is defined as a statistically significant change in the mean state of the climate or in its variability, persisting over an extended period (typically decades or more). These may be due to natural internal processes or external forcings, or to the persistence of anthropogenic variations in almosplere composition or land use.

They also refer to all climatic characteristics at a given location over time: warming or cooling (www.actu-environnement.com consultë le 21/10/2015).

These two dëfmitions are taken into account in our study.

Global warming: phënomëne of increase in the average temperature of the earth and oceans on a worldwide scale, the main cause of which is the increase in greenhouse gases in the earth, resulting from human activity (Fedele C., 2010).

It is also the increase in the average temperature at the surface of the planet. It is due to greenhouse gases released by human activities (Industry, transport, agriculture, ...) and piëgës in I'titmospliere (www.climati.e-monsite.com consulted on 21/10/2015).

These two definitions correspond well to the concerns of our study.

Microclimate: climate resulting from the effect of human action (planting and construction). It is therefore mainly "local". Urban microclimates are complex because they depend on the morphology of blocks and public spaces, and there are many factors involved (obstacles to thermal radiative fields and wind action). (Fedele C., 2010).

It also refers to all the mëtëorological conditions of a small geographical area which differ from the general climate of the area considered. These local specificities are due in gënëral to local topographic, gëological and hydrological cara^ristics (www.futura-sciences.com , consulted on 21/10/2015).

Evapotranspiration: the quantity of water transferred from the soil to the atmosphere by evaporation (passage of water from the liquid state to that of water vapour) at soil level, and by transpiration (elimination of water vapour in excës) by plants. (Fedele C., 2010).

It is also a process by which living organisms (especially plants) lose water in the form of vapour. It is therefore a loss of water due to two phënomënes: evaporation of water from the soil and plants and transpiration of plants. This phënomëne is at l'origine de la montëe de la serve dans les vaisseaux (www.aquaportail.com consulted on 21/10/2015).

Greenhouse effect: a natural thermal phenomenon which, for a given absorption of energy, gives the body receiving it a much higher surface temperature. Some of the sun's rays reach the ground, emitting thermal radiation that is absorbed by greenhouse gases and heats up Ihlinospbere and the ground (Fedele C., 2010).

It is also a phenomenon that heats up the surface of the Earth and the lower layers of the Earth's crust, due to the fact that certain gases in the atmosphere absorb and reflect back some of the infrared radiation emitted by the Earth, which makes up the solar radiation that the Earth itself absorbs. (www.notre-planete.info consultb le 22/10/2015).

Partial conclusion

There are three types of UHI that interact because of heat and mass exchanges.

These are :

- Underground UHI, which tends to increase ground temperature in urban areas by conduction;
- the surface UHI, which is characterised by the difference in surface temperatures between
- ities and rural areas are both pbriphoric, with urban surfaces warmer than those in rural areas;
- and the atmospheric UHI, which relates to air temperature (Leconte, 2014).

The three types of UHI interact because of heat and mass exchanges. The second type, surface UHI, is well suited to the concerns of our study. The effect of the UHI is that temperatures are not reduced sufficiently at night, which prolongs heatwaves because the temperature remains high at night and accumulates during the day, after which the cycle starts all over again. In addition, the characteristics of the urban form, the extent of the vegetation cover, the properties of the building materials and the scale of anthropogenic heat emissions are among the human factors involved in the formation of UHIs (Figure 2). Thus, it is all the factors combined together that make up the UHI. The commune of Porto-Novo and its surroundings are affected by the UHI through all of these factors (Figure 2).

NATURAL FACTORS		HUMAN FACTORS	
Geographic		Urban form	Plant cover
- **Climate,** - **Seasons** -**Topographic**	**ICU**	Modification : - Sun exposure - Speed and flow of wind - Radiative cooling - Geometric canyons urban	Decrease : - Green spaces, - Wetlands, - Permeable floors ; - Sweating plants and evaporation of water from the soil.
Weather conditions		Building materials	Anthropogenic heat
- **Sunshine** -**Precipitation,** -**Winds**		Solar energy: -absorbed and stored (radiative properties) - stored (thermal properties) High heat retention, albedo (dark paint, asphalt, roof, walls, insulation, etc.)	Broadcasts by : - Buildings, - Transport, - Industry; - Air conditioning; - Motorbikes, vehicles
Spacing homes apart **Use light-coloured materials**	**Attenuation ICU :** **GREEN POLICY**	Creation of a green grid	Greening towns (roofs, streets, houses, gardens, parks, etc.)

Figure 2: Conceptual diagram illustrating the factors that contribute to the formation of UHIs and the possible solutions.

CHAPTER II: PRESENTATION OF THE RESEARCH ENVIRONMENT

In this chapter, it is prëseпlë the study environment through the дёодгарЫдие situation and the physical and socio-economic earaeteristics.

2.1. Geographical location

The research environment encompasses 1 gëographic space structured by the city of Porto-Novo. Thus, it consists of this demiere and the satellite communes of Лкрго-M188ёгёlё, Avrankou, Sëmë-Podji, Adjarra and les Aguegues. It lies between 6°22'0" and 6°38'20" North latitude and between 2°30'0" and 2°41'15" East longitude (PDC 2015). Porto-Novo, the central city to which the other communes are attached, is the political capital of the Rëpublique du Bëпт and is located 13 кПотёlгез from the Atlantic ocëan then about 30 кПотёlгез from Cotonou. The research environment is structured into 31 arrondissements, 273 villages/town quarters and covers an area of 642 km^2 . Figure 3 presents the geographical situation of the Commune of Porto-Novo and the satellite communes.

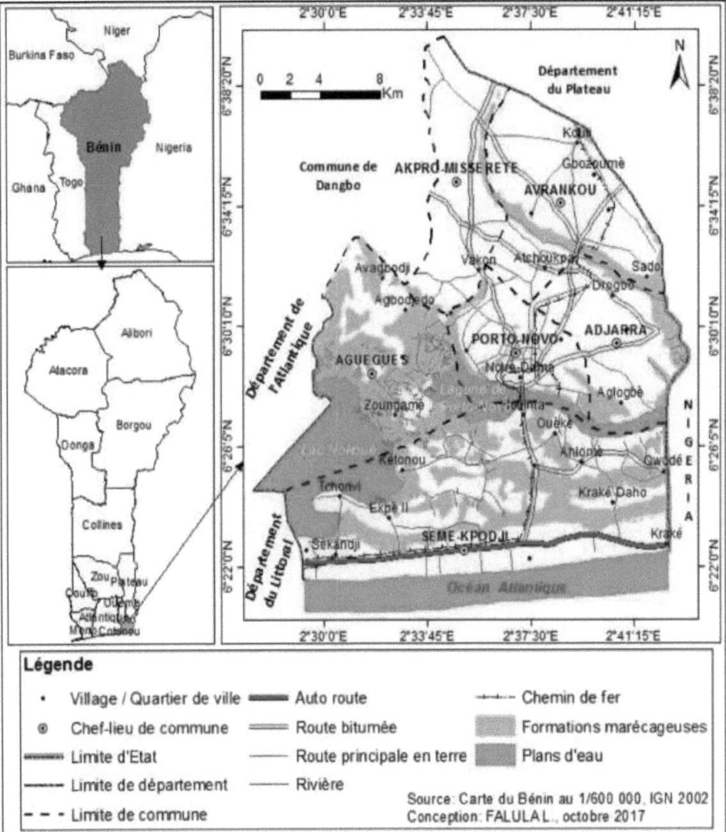

Figure 3: Situation дёодгарЫдие of the Commune of Porto-Novo and its surroundings.

2.2. Physical characteristics

The ëlëments taken into account here are relief, morphology, climatic factors, 1 hydrology, soils and gëological landscape and vëgëtation.

2.2.1. Imorphological components

Morphological components involve the relief in place, which undergoes changes at any time. In the research area, apart from Sëmë-Podji and les Aguegues, the communes of Porto-Novo, Акрго-M188ёгёlё, Avrankou, Adjarra are made up of plateaux. The relief is very uneven; with an altitude of less than 60 m, the relief presents notches in places.

There is a valley between the Commune of Adjarra and the Rëpublique Fëdërale of Nigeria. As for the Commune of Sëmë-Podji, it lies on a plain coPёre encased in a complex of bodies of water (océan Atlantique, lagune de Porto-Novo, fleuve Оиётё and lac Nokoue). The very low relief varies in places between 0 and around 6 m in altitude. It is mostly сотрозё of marecages, fine sands unsuitable for agricultural activities and bodies of water. The arable area makes up 39.5% of the total area of the commune. The Commune of Les Aguegues is situated on a relief that is characterised by 2 levels of altitude, gradually rising from south to north. The Commune's plains are made up of low-lying floodplains, crossed by the Оиётё river and its tributaries, the banks of which form bulges of land where people live. This is the vëritable low valley of l'Оиётё (PDC Porto-Novo, 2015). Figure 4 presents the morphological components of the study area.

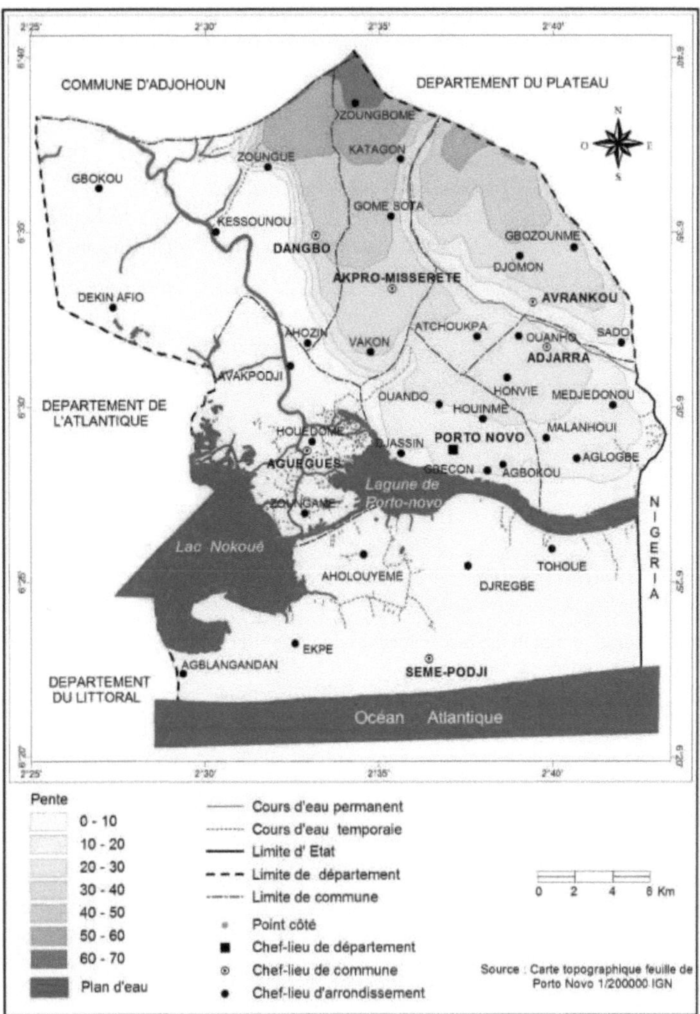

Figure 4: Topographical components of the study area

Figure 4 shows the morphological components of the study area. Analysis of this figure reveals that the research area is composed of plateaux and floodplains. The slopes vary between 0-10%, 10-20%, 20-30%, 30-40%, 40-50%, 50-60% and 60-70%. This means that slopes vary from 0.10 metres to 0.70 metres in the study area. In fact, if you travel one metre horizontally on a road and it rises by 0.10 metres, you will only be dealing with a road with a gradient of 0.10 metres, i.e. 10%. If it rises by 0.20 metres for each metre travelled horizontally, its gradient will be 20% and so on.

The gentle slopes are located to the south, Test and north-east (PDC, 2015).

2.2.2. Climatic factors in the study area

The Communes of Porto-Novo, Лкрро-M188ёrёlё, Adjarra, Sёmё-podji, Avrankou and

17

d'Aguegues share with the Dëpartement of 1 Oиётё a subëquatorial climate strongly influenced by the sudano-gui^en regime which ciiracterise the whole south Bëпт. This regime is characterised by a double alternation of dry and rainy seasons:
-    a long rainy season (April-July)
-    a short, dry season (August-September)
-    a short rainy season (October-November)
-    une grande saison seclie (deceeiibre-nuirs). (Adam and Boko, 1993).

Rainfall maxima are generally obtained in June for the long rainy season and in September for the short rainy season (ASECNA, 2004). Figure 5 presents the rainfall amounts in mm of the research environment over the last 30 uniëres (1985- 2015).

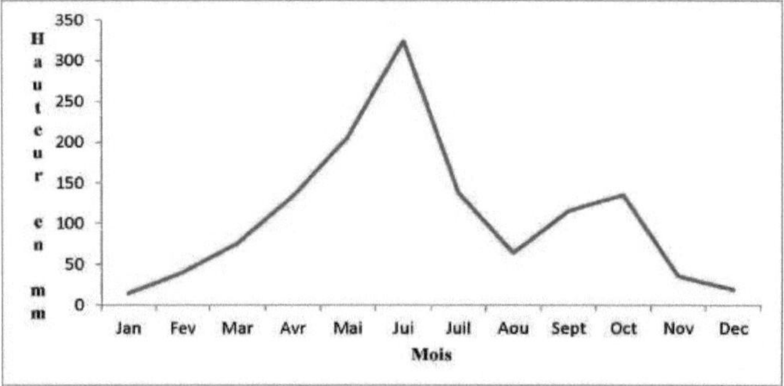

Figure 5: Rainfall in mm in the research environment **Source**: ASECNA data processing, 2017.

From the analysis of this figure, it can be seen that the peaks of the обзегуёез frequencies correspond to the months of June and September. This is the period with the highest rainfall.

In addition, the alternation of hot and humid periods and the proximity of the sea (around 15 km from the coast) are factors that influence tempëratures. For the two hottest periods, the average monthly tempërature is 30.5°C; it is 25°C for the coldest periods. Humidity rises to 80% in the rainy season and drops to 65% in the dry season.

The commune of Porto-Novo and the satellite communes are subject to the influence of two main winds of average intensity (2 to 4 mëtres/second on average); these are the monsoon, which becomes more intense between January and July, and the harmattan, a cool, dry wind that is active from December to January. The monsoon blows from the south to the north of the country and the harmattan from the north to the south.

The average air tempëratures in the city over the last 30 years are presented in Table I (Appendix I) and interpreted in Figure 6.

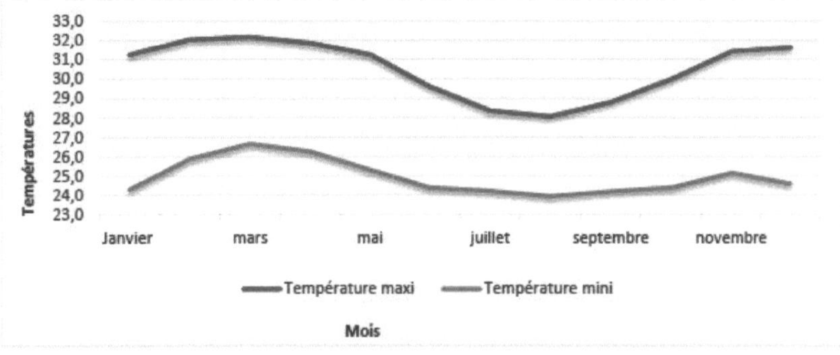

Figure 6: Average monthly tempëratures over the përiod 1985-2015 in Porto-Novo and the surrounding area

Source: ASECNA data processing, 2017

The figure shows that the hottest months of the year in terms of maximum temperatures are January, February, March, April, May, November and December, with an average value of 30.5°C. In contrast, the warmest months in terms of minimum temperatures are February, March, April, May and November, with an average value of 25°C. On the other hand, for minimum tempëratures the hottest months are: fëvrier, March, April, May and November with an average value of 25 °C.

2.2.3. Hydrographic network

The hydrographic system is essentially made up of the Porto-Novo lagoon, which is relatively shallow (between 2 and 4 metres deep), as well as the Zounvi, Donoukin and Vakon rivers. This lagoon is part of the hydrographic network of the Oнëтë and So rivers. Elie dëverse dans l'ocëan Atlantique par le chenal de Lagos. The depth of the water table varies according to the topography near the lagoon and in the dëpressions it is less than three (3) metres, while on the plateau it is around 15 metres. As for Akpro-Misserete, it is made up of ten (10) kilometres of waterways covering four rivers and a few marigots. There are also marecages and shallows suitable for fish farming in several arrondissements (Vakon, Katagon, Gomë-Sota and M188ërëlë). The Avrankou commune is criss-crossed by a hydrographic network made up of streams and lowlands. The lowlands surround almost the entire commune and cover an area of around 16 km^2 . They are crossed by 46km of watercourses, most of which are navigable. These rivers are Houssoutokpa, Atchoukpa tokpa, Gbokouso tokpa, Danmë kpossou tokpa, Wamon tokpa, Sado tokpa, Agoumanya, Sogbo, Adogba and Tokpa agua. Sëmë-Kpodji, located between the sea, lake and lagoon complex, Sëmë-Podji bënëflcie d'un reseau hydrographique favorable aux t-ictivites de peche. This is the Cotonou lagoon, which widens to form Lake Nokorie (14,000 ha). It communicates via the Toclie canal with the Porto-Novo lagoon, which extends as far east as Lagos in Nigeria, creating a kind of freshwater reservoir (PDC Porto-Novo, 2015).

The hydrographical network of the commune of Adjarra is made up of the Porto-Novo lagoon in the south and the Aguidi river in the north-east. The bed of this river is occupied by raffia palm trees. It appears to be a string of low-lying areas used by local people to obtain water from a number of sources (the marigots of Do, Tchakou, Sëmë, Mëdëdjonou, Djavi, Adjina, Adjarra etc.) and transverse canals. The water bodies are poorly maintained. They are invaded by vëgëtaux and dëchets which make it difficult for them to flow. As a result, they have become poor and are gradually filling in. As far as the Aguegues are concerned, the main watercourse running through this commune is the 1Юиёгё delta. This commune forms the bulge of land and the vast plains of marshy lowlands that sëparent the Porto-Novo lagoon and Lake Nokorie. The large Totche canal is its dëmarcation line to the south with the Commune of Sëmë-Podji (PDC Porto-Novo, 2015). Figure 6 illustrates the hydrography of the study area.

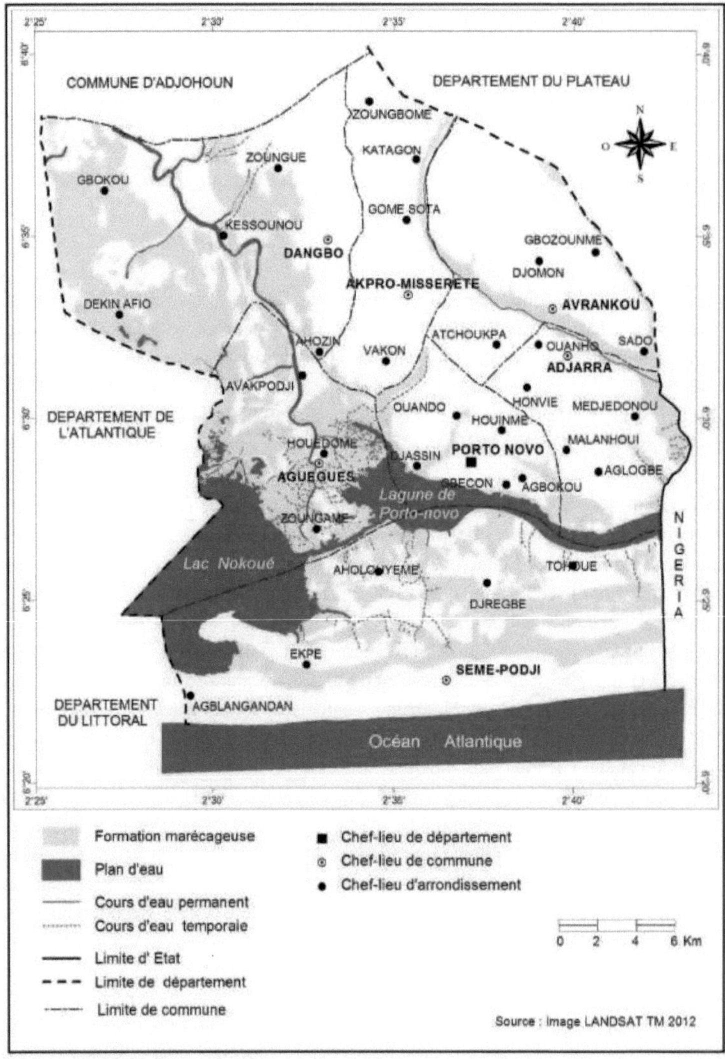

Figure 7: Hydrographic network in the study area

Analysis of this figure shows that the research area contains permanent and temporary watercourses, wetland formations and water bodies. These various watercourses and water bodies offer a wide range of activities, from market gardening to fishing and river and lagoon

20

transport. Around these watercourses and bodies of water, there is a vëgëtale formation made up of herbaceous plants and shrubs. These are places that are home to diverse ëcosystëmes. These plant species help to clean up the environment by capturing carbon dioxide in suspension.

2.2.4. Soil and geological landscape

In the research area, soils vary from one commune to another. In the commune of Porto-Novo, two types of soil can be distinguished: leached soils on the plateau stretching from Djlado to Tokpota via Dowa, Houinmd, Hounsouko, Djdgan and Attakd. These soils consist mainly of a layer of reddish-ochre clay and sand (terre de barre) alternating with a stratification of yellow, white and clayey sands. These soils are leached and impoverished by the intensive cultivation of oil palms, coconuts and various food crops such as maize, manioc, groundnuts and nidbd. Hydromorphic soils are found mainly in the low-lying areas along the Porto-Novo lagoon (Profil Environnemental de Porto-Novo, 2001).

Moreover, according to the classification drawn up as part of the National Programme to Combat Desertification, Porto-Novo is located in the zone of highly degraded soils with high demographic pressure (Environmental Profile of Porto-Novo, 2001). In these areas of barren land, often densely populated, the fragmentation of land into very small holdings due to the inheritance system and social distribution mechanisms is not conducive to the practice of fallowing. A decline in soil fertility is thus due to the erosion of surface formations. In the town itself, the soil consists of a mixture of clay and sand to a depth of around 17 m, characterised by good drainage and high cohesion. Unlike in Cotonou, these soils are relatively impermeable, with coefficients of the order of 10-7 m/s to 10-5 m/s (Environmental Profile of Porto-Novo, 2001).

The Communes of Akpro-Missdretd and Adjarra have three (03) types of soil: ferralitic soils, red in colour and with a sandy-clay texture (bar soils), which cover around 80% of the total area of these Communes. The light brown, sandy-textured soils, which are easy to work, are located on the edges of low-lying wetlands, in closed depressions, and the hydromorphic clay soils, rich in organic matter, are located in flood-prone areas (PDC, 2015).

The Commune of Avrankou has ferruginous soils formed on the terminal continental. They are deep and easy to work. They cover more than 80% of the Commune's soil. In some places, it is used as a red earth quarry for road building and for making stabilised earth bricks. The hydromorphic soils are very localised in the wetlands of the Commune, particularly in the marecageous zone; kaolin is present in the arrondissement of Atchoukpa, and peat is present in large quantities along the marecageous zone. Because of its combustible nature, it is an important but unexploited resource for the commune (PDC, 2015).

Due to its topographical position, the commune of Sëmë-Podji only has soils that are essentially the result of leaching or sëdimentation. They are mostly hydromorphic and very poor in ëlë nutrients and organic matërials, particularly in base, nitrogen and phosphorus, but rich in silicon dioxide with some ëlë elements of tropical ferruginous soils. Overall, a distinction can be made between hydromorphic soils in this Commune that are not very ëyolиёз and therefore poor, formed on sea sand, hydromorphic soils with Gley that are moderately organic, moist and richer, formed on alluvial lagoon matëriaux, lesives soils with a podzolic tendency formed on quaternary and pseudo-gley soils formed on sandy-clayey matëriaux As a result, very few soils are favourable or marginally suitable for food productionx On the other hand, they are apparently favourable for oil palms, coconut palms and sugar cane, which develop well there (PDC, 2015).

The commune of Les Aguegues has hydromorphic black clay soils suitable for agriculture.

These soils receive alluvial deposits annually when the river Ouëtë floods, which maintain its fertility. This Commune abounds with several canreres of river sand, black pottery clay and agricultural lowlands. Exploitation of these resources is not yet well organised, although they are an asset for the local population in terms of building houses and structures, handicraft pottery, off-season production and so on. Figure 7 shows the pëdological formations in the study area.

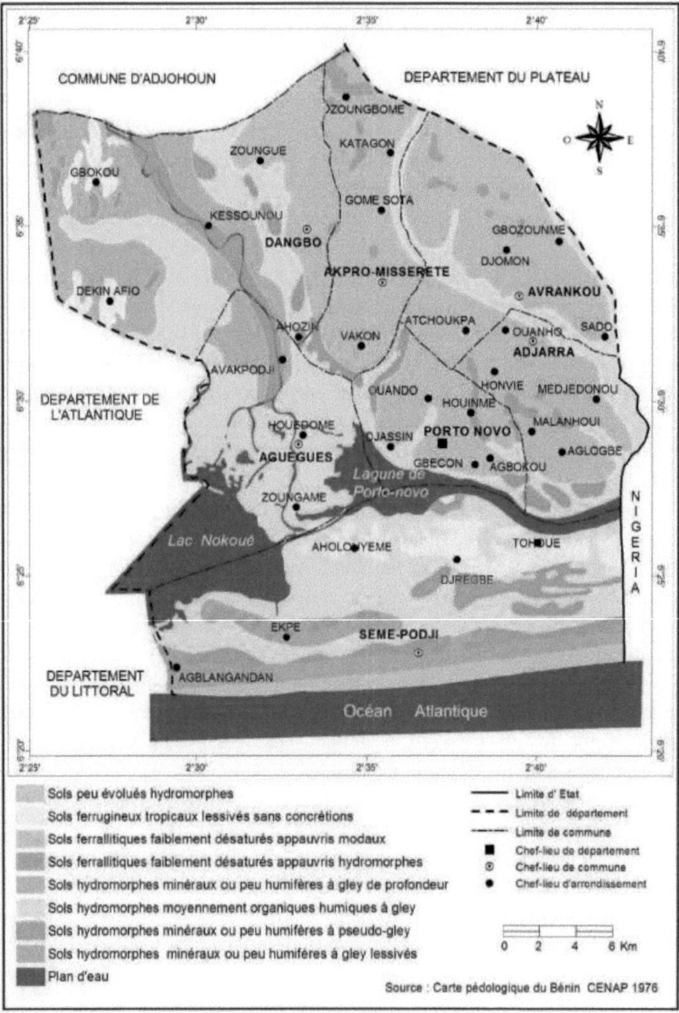

Figure 8: Pëdological formations in the study area

Analysis of figure 8 shows that in this zone, there are gënërally ferralitic soils, ferruginous soils and hydromorphic soils and their variants. On these soils grow a vëgëtation depending on its night components.

2.2.5. Plant training

La yëдë!айоп du milieu de recherche a etc largement modifree par I'activite humaine. Elie is сотрозёе of grassy savannah which is observed today in the extension zones whose 1 occupation évolue very rapidly due to the opёрайопз of housing estates which dё develop there. In places, it is sparseёe. Elie is dominated by oil palm (*Elaesis guineensis*). On the hydromorphic soils of the lowlands, an atypical vegetation of ferns, brambles and aquatic plants grows, while along the riverbanks there are a few species of forest under which the riverside populations grow food crops and market gardens. On the margins of the marshes, the more varied vegetation consists of raffia palm, bamboo, iburrager and other water-loving species. In the sandy zone by a weakly rooted herbaceous carpet, vast coconut groves, filao, 1 eucalyptus, and 1 acacia. The marecageous zones: by a few clumps of Andropogon gayanus, isotetic stands of bramble (Borassus aethiopum) and cyperaceae (PDC, 2015).

2.3. Human and economic characteristics

2.3.1. Demographic data

Dans la zone d'étude eonstituee par les Communes de Porto-Novo, Акро-M188ёрёlё, Adjarra, Sёmё-kpodji, Avrankou et d'Aguegues, des résultats du recensement 1979 il est dёnombrё respectivement des effectifs de la population de 133168 habitants, 39291 habitants, 34074 habitants, 5233 habitants, 50016 habitants et 14895 habitants. In 1992 these populations increased to 179138 inhabitants, 52 885 inhabitants, 46 427 inhabitants, 65016 inhabitants, 68 503 inhabitants and 21 333 inhabitants respectively. In 2002 the population had risen to 223,552, 72,652, 60,112, 115,238, 80,402 and 26,650 respectively. Population numbers in the study area grew in 2013 to 264,320 inhabitants, 127,249 inhabitants, 97,424 inhabitants, 222,701 inhabitants, 128,050 inhabitants and 44,562 inhabitants respectively. Figure 8 shows population trends in the research area from 1979 to 2013.

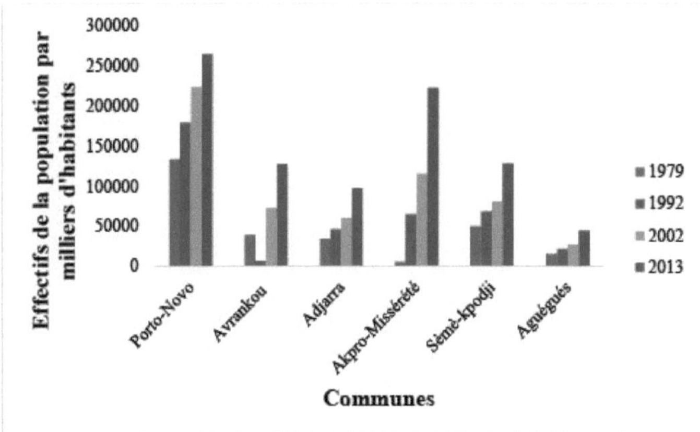

Figure 9: Population trends in the research area from 1979 to 2013

Looking at this figure, it emerges that the population of the area is experiencing an increasing ёyo1ийоп from 1979 to 2013. This strong growth poses problems of space management, especially in urban areas, where everyone is seeking to live in the city, leaving rural areas as a result of the depletion of agricultural land. As a result, new activities require people to move, mainly using individual means of transport, which generates atmospheric and noise pollution through engine exhaust fumes. Added to this are the gases released by the few industries located there. The sanitation problem is not overlooked, with liquid and solid waste disrupting people's lives through escaping gases, resulting in illness.

2.3.2. Demographic trends and projections

Two main periods should be considered in the demographic growth of the city of Porto-Novo, around which the other Communes are grafted. Between 1961 and 1979, the population of Porto-Novo rose from around 61,000 to 133,163, with an average annual growth rate of just over 4%.

During the intercensal period (1979-1992), the population rose from 133,168 to 179,138, with an average growth rate of 2.3%, i.e. an average of 3,584 new inhabitants per year. This fall in the growth rate is essentially due to the decline in the migratory contribution; the migratory balance fell from 79 to less than 2,100 inhabitants over the period as a result of the investments made by the State in the agricultural development sector from the 1980s onwards. On the other hand, the average natural growth rate rose slightly from 2.9% to 3.4%. Taking into account the 2013 population with a growth rate of 3.69% for the département of l'Оиёте and the formula (RGPH, 2013).

The dëmographic projection of the research area in 2025, based on the calculations, gives the following results: 394,799 inhabitants, 190,064 inhabitants, 145,517 inhabitants, 332636 inhabitants and 191,261 inhabitants, 66,560 inhabitants. Figure 9 shows the population of the research area in 2025.

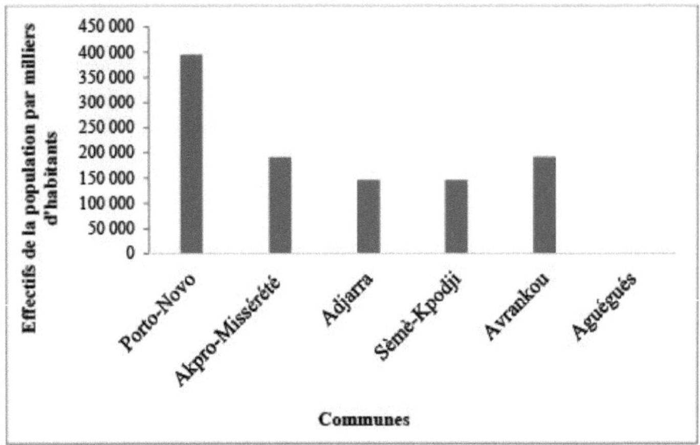

Figure 10: Population of the research area by 2025

Analysis of this figure shows that the population of the research area varies from one commune to another. On the one hand, this growth would be natural, and on the other, it would be related to internal migration, which pushes young people and certain adults to leave the rural environment for several reasons, including soil impoverishment, social problems and improved living conditions.

2.3.3. Ethnic groups

Today, there is a mosaic of ethnic groups living together in the research area. These include the Goun, the Yoruba, the Toffin and the Оиёте. Thus, the Goun and Fon are in the majority (66%), followed by the Yoruba (25%), and the Adja, Mina and Toffin (4%). Other ethnic groups include the Bariba, Dendi, Yom-Lokpa, Otamari and Peulh (5%). The Commune of Akpro- Missërëtë is populated mainly by Tori (97.4%) and Yoruba and related groups (1.3%), and also includes Yoruba and Ibo from Nigeria. As for the Commune of Adjarra, it is inhabited by Goun (83%) and Yoruba (8.2%), Adjarra, Setto, Toli, Goun and Yoruba are the languages spoken there. The Commune of Avrankou is populated mainly by Tori and apparentës (93.7%) and Yoruba and apparentës (3.5%). In the Commune of Sëmë, there are the Fon and apparentds, who represent 46.19% of the commune's inhabitants, and the Adja and apparentds, 37%. The other groups are in the minority; these are the Yoruba who represent only the 7.25% In the Commune of Aguegues, Toffin and the Онёгё constitute the two majority ethnic groups that cohabit (PDC, 2015).

This ethnic mix is also at the root of the diversity of economic activities in the town. The Yorouba traders have developed their commercial activities, while the Goun and Fon are heavily involved in agriculture and transport. As for the other ethnic groups, they are to be found in the provision of services, in refreshment stands and restaurants and in miscellaneous.

2.3.4. Religion

The spiritual life of the research area is animated by several religions. Each of them preaches a culture of peace, mutual tolerance and local and national cohesion. Three categories of religion can be distinguished:

1. traditional religion (29.20%): Traditional religions are centred around vodun, tron, zangbdto, d'oro, etc. The demands of their rites and rituals are conducive to the protection of the sacred forests that house their convents.

2. the Christian religion (45.70%): this includes Evangelical churches, Catholics, Protestants, Christian celestials, etc.

3. l'Islam (25.10%)

The cultural identity specific to this research area is based on the triptych of ancestral beliefs and religious syncretism constituted by the belief in a supreme God, creator of the universe, and the cult of ancestors known as "Vodoun" in Goun or "Orisha" in Yoruba.

2.3.1.5. Town planning

Since the 1960s, urban development in the city of Porto-Novo has extended to the neighbourhoods around the outer boulevard, such as Kanddvid, Houinmd, Djaguidi, Founfoun and Avakpa. At present, urbanisation is taking place in neighbourhoods such as Dowa, Akonabod, Djdgan-Kpdvi, Djdgan-Daho, Gbodjd, Louho, etc.

The city of Porto-Novo covers 5,213 ha with an urbanisable area of 4,415 ha and a spatial growth rate of 2.6% (PDC, 2015).

The town contains 430 ha of low-lying land. The remainder is divided between the districts as follows: Ouando (1.423 ha), Houinmd (366 ha), Hounsouko (426 ha), Djegan Daho (976 ha), Ilefid (75 ha), Akron (71 ha), Avassa (41 ha), Oganla (60 ha), Houezounmd (21 ha), Zdbou (16 ha), Ahouantikome (35 ha), Degue gare (58 ha), Foun-Foun (215 ha), Djassin (281 ha), Attake (466 ha), Bas-fonds (430 ha) (PDC, 2015)

The servicing of the central districts is good. This is not the case in the new districts where l'eau et Pë^скгас^ë cover only the primaries and secondaries streets. Land management in a large city like Porto-Novo is of nëcessitë in order to better control subdivisions but also mutations of rights over plots. To this end, the urban land register (RFU) has been set up.

2.3.1.6. Habitat

o Type of habitat and materials used

Housing in the city of Porto-Novo and the surrounding Communes is made up of traditional houses built of earth, bamboo, brick and plots of land, and modern houses built of brick, sheet metal and slabs. The dwellings are organised in family concessions, especially in the old districts of Porto-Novo's arrondissements 1 and 2, while everywhere else there are y family concessions.

plots and villas.

Photo 1: Porto-Novo seen from the sky (30 тё1гез resolution).
Source: IGN France, 2018

Photo 2: Porto-Novo from the air
Source: Google MAP, 2022

One of the particularities of the city of Porto-Novo resides in the contrast presented by the urban landscape. Indeed, y distinguish several types of architecture that harmonize to make the beautyë of the city (PDC Porto-Novo, 2016) :

1.   traditional architecture: it marks the ancient core occиpë by the Goun and Yoruba family concessions centres around the royal palace "Honmë". This architecture can also be observed in its përiphëric Communes. This architecture is characterised by the vëtuscitë of the building materials used, even though some traditional buildings have been refurbished.

2.   1 colonial architecture: this can be seen in the colonial administrative zone, with

26

monumental buildings used as workplaces and administrative residences.

Photo 3: Old Central Mosque in Porto-Novo
Photography: L. FALOLOU, December 2019

3. 1 Afro-Bresilian type architecture: it is 1ocaH3ёe in the junction space between the old core and the colonial administrative zone to the west of the city. This is the model yёЫси１ё by the freed slaves and inspired by Brazilian or Portuguese-style buildings. The buildings are imposing and marked by dёcorative motifs. The central mosque in Porto-Novo is one of the most representative prototypes of this model.

4. 1 Religious architecture: This is typical of convent temples, ё churches and mosques (inspired by Portuguese and Middle Eastern architecture).

Photo 4: Notre Dame Cathedral in Porto-Novo
Photography: L. FALOLOU, December 2021

Photo 5 : New Central Mosque next to the old one in Porto-Novo
Shot by L. FALOLOU, December 2021

5.   Contemporary architecture in the "new districts" on the outskirts of old Porto-Novo.

In the commune of Akpro-Missdretd, traditional and semi-modern dwellings characterise the construction methods. Overall, housing has not changed much, even in areas where the land is already parcelled out and belongs to individuals or communities. In Avrankou, the transport situation is almost identical. The urban centre looks like an old town with old buildings. A mixture of mud huts and permanent houses built by the local authorities make up the overall decor of the town. Prestigious buildings (modern housing) are ёпдёз in places but in very limited numbers. However, there are currently major construction sites сопзёсийГз to subdivision work which augurs for soon a new face of the city which houses most of the ёadministrative facilities of the Commune.

Housing in the commune of Sёmё-Podji is predominantly groupёnёral. The highest concentrations of people are found around the arrondissement administrative centres. Most of the dwellings are made of durable materials (cement, sheet metal, iron, etc.) and sometimes of precarious materials (palm ribs, clays, raffia ribs, etc.). Their architecture is recent.

There are three types of habitat in the Commune of Adjarra:

-    Traditional dwelling: Bare earth or cement mortar rendered construction covered in straw or roofing inside an unfenced plot.

Photo 6: Traditional housing in Adjarra *Shot: L. FALOLOU, December 2017*

-    Semi-modern dwelling: Bare earth or cement mortar rendered construction covered with insulated roofing or inside a fenced compound with a gate.

Photo 7: Semi-modern housing in Adjarra
Photography: L. FALOLOU, December 2017

\-   Habitat moderne : Construction en aддlотёгёз de ciment et couvert en tole, tuile ou dalle en Ьёlоп a l'intёrieur d'une concession cloturee.

Photo 8: Modern housing in Adjarra
Photography: L. FALOLOU, December 2017

Overall, 1 habitat has not much ёуоlиё even in areas where land is dёja parcellisёes and owned by individuals or collectives.

The urban centre has the look of an old town with old buildings. A mixture of banco huts and solid houses separated by the vёgёtations forms the overall dёcor of the town. Modem buildings are ёrigёs in places but in very limited numbers. However, there are currently a number of large construction sites consёcutifs to the subdivision work, which suggests that the town, which houses most of the Commune's administrative facilities, will soon have a new face.

As for the commune of Aguegues, we find traditional dwellings: houses on stilts and also houses made of dёfmitive matёriaux either on stilts (in Ьёlоп) or built directly on the ground if the texture allows.

2.3.2. Economic characteristics

In the research area, economic characteristics vary from one commune to another. The local economy of Porto-Novo is essentially based on the informal sector. The formal sector is not developed. This situation is helped by the permёabilitё of the Bёnino-Nigёriane borders. The city has not specialised in any particular economic activity. However, Porto-Novo remains the тёкорок of large Bёninois merchants with relatively large turnovers.

As Figure 10 below shows, 47.28% of the sons of Porto-Novo are involved in the tertiary sector. They are mainly involved in trade, the development of which is supported by their large neighbour Nigeria. The second sector that mobilises the population of Porto-Novo is manufacturing (26.29%). There are only a limited number of industrial businesses registered in Porto-Novo (21). On the other hand, the craft industry is a major ëlëment of the 8pëe1:йc11ë town, particularly in terms of employment and income. As for the primary sector, it is non-existent (2.65% of the population), which shows that the population of Porto-Novo is not agricultural.

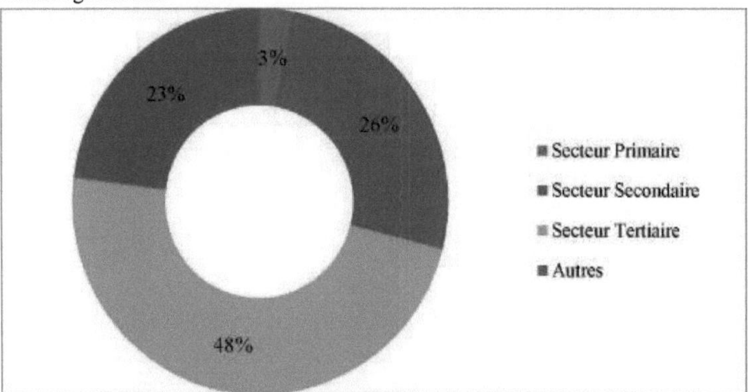

Figure 11: Rëpartition of the population of Porto-Novo by economic sector *Source: Plan de Developpement Commiunale (PDC) de Porto-Novo*

The economically active population is largely dominated by women, who run more than 56% of the businesses surveyed, particularly in the retail sector. The working population is young, and 54% of heads of commercial and service businesses are under the age of 30. The "modern" circuit in which the Yoruba are found in imijonte developed with the oil boom in Nigëria in 1973. The geographical proximity of this country and the ethnic links encourage more or less tegal ëchanges between traders through, on the one hand, the development of a sector described as informal and, on the other, the expansion of the urban area. The economic activities of the Commune of Akpro-Missërëtë are essentially based on the informal sector, with 52.84% of the population involved in the tertiary sector. They are mainly involved in trade, the development of which is коуопзë by the big neighbour Nigëria. The second sector that mobilises the population of Akpro-Missërëtë is the primary sector (31.43% of the population). Agriculture is the dominant activity in this sector. There are no registered industrial enterprises in the municipality of Akpro- M188ërë1ë. The 7.83% of the population in this sector are mainly involved in handicrafts (Figure 10).

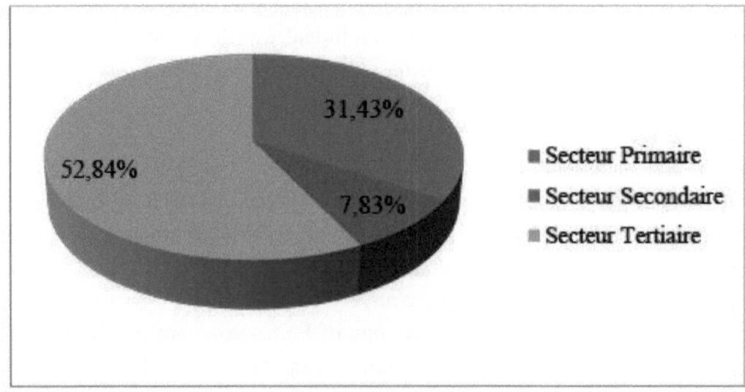

Figure 12: Rëpartition of the population of Akpro-Missdrdtd by economic sector *Source: Akpro-Misserete Commiunale Development Plan (PDC).*

The tertiary sector (52.84%) takes precedence over the secondary sector (7.83%) and the primary sector (31.43%). The urban phenomenon engulfs 52.84% of this population, which is subject to various related consëquences.

The local economy of Avrankou is essentially based on the informal sector, favoured by the permedability of the Bënino-Nigëriane borders. According to the figure above, the population of Avrankou invests 53.51% in the tertiary sector. The inhabitants are mainly involved in commerce, the development of which is favoured by their large neighbour, Nigëria. The other activities in this sector are catering and transport. The primary sector is the second sector that mobilises the population of Avrankou. In this sector, 21.20% of the population are involved in agriculture, hunting and fishing. The third sector that mobilises the population of Avrankou is manufacturing and construction (15.19%). There are no industrial enterprises located or registered in Avrankou. Figure 13 shows the distribution of the population of Avrankou by sector.

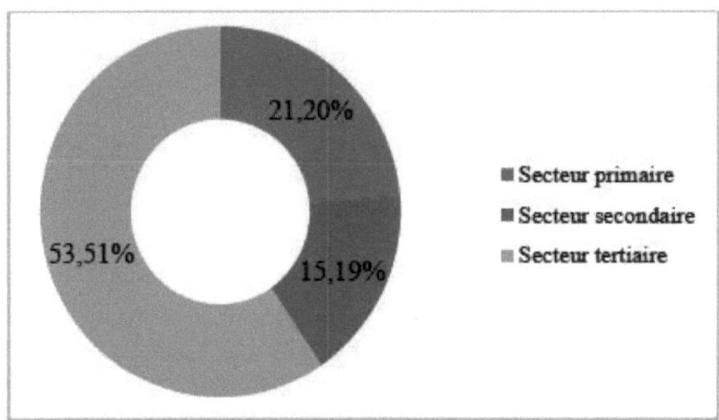

Figure 13: Rëpartition of Avrankou's population by ëconomic sector.
Source : Avrankou Community Development Plan (PDC)

Analysis of this figure shows that the tertiary sector accounts for most of the population of this municipality, followed by the primary and secondary sectors. This situation gives rise to problems of gas management of all kinds, which contribute to the formation of heat islands, especially in the urban environment represented by the central districts.

The local economy in Adjarra is essentially based on the informal sector, favoured by the permed nature of the borders between Benin and Niger. According to Figure 14 below, 56.91% of the population of Adjarra work in the tertiary sector. They are mainly involved in trade, the development of which is encouraged by their large neighbour, Nigeria. The second most important sector for the population of Adjarra is manufacturing (19.17%). There are no industrial companies established or registered in the commune of Adjarra. On the other hand, the craft industry is a major factor in the specificity of Adjarra, particularly in terms of employment and income. As for the primary sector, it is practised by 13.33% of the population, which shows that the population of Adjarra is less and less agricultural.

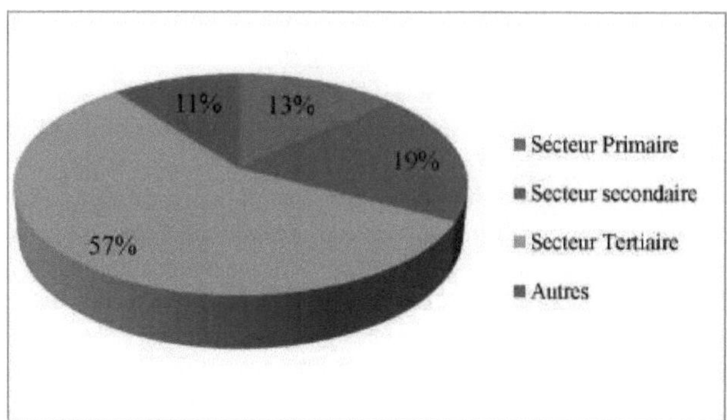

Figure 14: Rëpartition of the population of the Commune of Adjarra by sector ёсопот!дие

Source: Plan de Developpement Commiunale (PDC) d'Adjarra.

Analysis of this figure shows that in this Commune, the tertiary sector accounts for 57% of the population, compared with 19% for the secondary sector, 13% for the primary sector and 11% for the other sectors. This shows that Adjarra is a përiphëric town to the Commune of Porto-Novo. It contributes to the growth of its population through the services it provides, which is not without consëquences for the production of greenhouse gases from various sources such as transport, storage of household waste, etc.

2.3.2.1. Trade

Commerce is a key economic activity in the Commune of Porto-Novo. It employs 47.28% of the population and covers a wide range of products. These include hydrocarbons, manufactured products, pharmaceutical products, building materials, foodstuffs, beverages, cosmetics, agricultural products, livestock products and so on. These products come from Nigeria, Cotonou (via the port) and the rural areas of Porto-Novo. This activity is mainly carried out by women.

Commerce plays a key role in Adjarra's local economy. It mobilises 47.61% of the population and involves a variety of products. These include hydrocarbons and manufactured products, mainly from Nigeria, agricultural products, livestock, handicrafts, processing and pharmaceuticals. Most of this activity is carried out by women, who are actively involved in petty trading. However, it should be noted that the Commune is also home to a number of large nationally-recognised traders.

In the Commune of Avrankou, trade is a popular activity due to its proximity to Nigeria. Les produits commerisës peuvent être гедгоирёз en trois саlёдопез qui sont les produits përoliers, les produits manufactures, les produits de transformation locale. These products are exposedës along the main arteres of the Commune.

In Sëmë-Kpodji, the trade sector is not very developed but is sufficiently diversified. It is mainly informal and carried out by small-scale, low-income traders involved in the wholesale and especially retail sale of harvested products (sugar cane, coconut, sweet potatoes, cassava, maize, rice, etc.), processed products (' sodabi' or 'wine') and foodstuffs.), processed products ("sodabi" or palm wine, "gari" or cassava flour, etc.) and manufactures (drinks, building materials, foodstuffs, etc.), petroleum products (petrol, diesel, рёкок, motor oil) and pharmaceuticals that come fraudulently from Nigeria.

2.3.2.2. Commercial infrastructure

Porto-Novo has eleven (11) markets, two (2) of which are of capital importance have recently гёпоуёз. These are the central market of Porto-Novo and that of Ouando. The Ouando food market, five кПотёксв from the city centre, is increasingly asserting itself as Porto-Novo's most important commercial centre with a regional character. As for the central market of Porto-Novo, apart from food products, it is more spëcialisë in manufactured products (preserves, fabrics, drinks etc.).

The commercial facilities in the Commune of Akpro-Missёrёtё are weak. It is based on the existence of a few shops, storage shops and sheds built in three functional steps, two of which are daily. These sheds are made of both durable and precarious materials. Most of these markets are not amёnagёs. There are, however, several marches that come alive every evening.

In Avrankou Commune, each arrondissement has at least two markets, including a night market. The main markets in the Commune are the Avrankou and Kouti markets. These markets are not built of durable materials, and the access roads are not improved. The Commune also has a number of hardware and sundries shops. These shops are located along the asphalt road and around the main steps.

The commercial infrastructure of the Commune of Adjarra is weak. It is based on the existence of a few shops, storage warehouses and sheds built in the Gbangni, Kpetou and Alladako marches (non-functional marclie). These sheds are made of both durable and precarious materials.

2.3.2.3. Transport infrastructure

The Commune of Porto-Novo regulates the transport of goods and people within its territorial jurisdiction. Analysis of urban roads, types of transport and transport ёequipment provides a better understanding of this sector in the city of Porto-Novo.

There are several types of urban road in Porto-Novo: paved roads, asphalt roads and earth roads. We should also mention the railway that crosses the town but is no longer in service. In recent years, several roads have been improved (paved or tarmac), particularly in the central districts. All of these roads are distributed according to their importance.

Goods and people are transported in Porto-Novo by motorbike taxis, individual wliicles and motorbikes, walking for urban transport and by уёькикз of all kinds for inter-urban transport. The motorbike taxi commonly called "zёmidjan" is the most widespread urban transport. Data on the number of zёmidjans, taxis and other means of transport are very poorly known and it is important to conduct ёtudes for the knowledge of the situation de dёpart.

However, it is collectively acknowledged that the city of Porto-Novo has more than 2,500 zёmidjans (motorbike taxis), which provide between 30 and 40% of urban transport. The other means of transport are motorbikes and private cars. Taxis and minibuses for transport within the city have disappeared since the emergence and development of motorbike taxis (zemidpins), which are better suited to the highly graded streets. The use of pirogues should also be noted in arrondissements 1 and 3. Bicycles and rickshaws are frequently ийНеёе for transporting certain types of goods. Transport to neighbouring towns is provided by light cars (5 to 6 seats) and buses and minibuses.

The ёquipements of transport are the auto-gares, and the embarcadёres. The city of Porto-Novo has several official and spontaiees stations. The most important official stations are:

- OUANDO station with 830 m^2 ;
- Dangbёklounon station with 5,000 m^2 ;
- Adjarra-docodji station with 2,500 m^2 ;
- Saint Pierre et Paul station officially closed recently;
- gare du Pont which will make way for the siёge de 1 Assembke Nationale.

Each auto-station is managed within the framework of co-management with the city, the aims of which are to involve the syndicate in the management of public transport and organise better coordination of public transport. The role of the road station committees is to collect loading fees and ensure зёсигкё, 1 order and cleanliness at road stations.

There are four (04) parking areas for buses and minibuses at Saint Pierre et Paul, Agbokou, Dëguëgan and Djassin. In addition, there are spontaneous car parks, including two (02) at Katchi and around Ouando. Despite the decline in river transport, there are three (03) embarcadëres in arrondissements 1 and 3.

In addition, several motorbike taxi pares and guard yёlo are сотр!ёз in the city. The motorbike taxi pares are very poorly organised given the strong тоЫШё of zёmidjans. The transport sector in the town of Porto-Novo is runë by several organisations and unions in collaboration with the Town Hall.

Inter-city transporters and zёmidjans are organised into several syndicates represented in the city of Porto-Novo.

Actions to amëliorate urban roads mainly concern trunk roads. In addition, the various proposals insist on the construction of certain гоиPёree stations. Il est ёgalement nécessaire de reorganiser le secteur et mettre en place un systëme de controle pour mieux connaitre la situation de diflerents acteurs du secteur ainsi que l^l élaboration d^a politique de gestion des transports urbains.

Transport in the commune of Akpro-Missёrёtё is essentially based on overland communication routes which serve as a support for means of transport that facilitate the dëplacement of goods and people. The Commune of Akpro-Missёrёtё is served by 26 km of national and interstate roads. Access to the capital is facilitated by a tarmac road in good condition. Road transport is provided by various means: bicycles, motorbikes, cars, buses and lorries. The most common form of road transport in the Commune is based on the use of motorbikes due to the phenomenon of motorbike taxis or "Zemidjan", of which Adjarra is the birthplace. Motorbike taxi drivers form a real professional body with a communal association.

The Commune has an asphalted interstate road, along which përiodic marches come to life, and a гоиPёre railway station located in the main town. In addition, the poor state of the access roads to the Commune's localities hinders the development of trade. In such a context, the transport of agricultural produce to consumer centres remains a serious obstacle to improving the living conditions of rural communities. Malgre l'importance relative des ёchanges mencs aux marches, les activites commerciales a Акрго-M1eeёrёlё sont dominoes par le trafic avec le Nigёria.

The commune of Avrankou has 8.46 kilometres of ЫШтёез roads and 66.44 kilometres of secondary roads and tracks that can be used at times to reach аддlотёгайоп8 and steps. There are several navigable access routes in the Commune. These include the following trongons: Houssoutokpa-Adjarra ; Atchoukpa tokpa-Djomon ; Gbokouso tokpa-Akpro Misserete ; Danmë kpossou tokpa- Akpro Misserete ; Wamon tokpa-Adjarra ; Sado tokpa-Djёgou ; Sado tokpa-N^ria ; Agoumanya- Kokoumonlou ; Sogbo-Ko Anangodo ; Adogba-Koadogba ; Tokpa agua-Atchoukpa. Waterways are navigated by pirogue. Motos appeks Zёmidjan taxis provide 80% of transport, but this sector is still informal.

In the Commune of Sёmё-Kpodji, the тоЫШё of men and goods is ensured thanks to a road network 395.79km long according to the Town Hall in its work entitled A la dёcouverture de la Commune de Sёmё-Podji 2 and composed of a motorway. The 24km-long Autoroute Inter Etat is made up of permanent main roads (20 trunk roads covering 85km including 9km ЫШтёз), permanent secondary roads (11 trunk roads covering 29km) and several seasonal roads and tracks. The means used in the commune for displacement are mainly уёЫеикв taxis and motorbike taxis, buses, mostly in transit, trucks. There y also cars, motorbikes and bicycles for prices. On the waterways it is pirogues, boats that are usedëes.

Transport is essentially based on two communication routes dedicated to the movement of goods and people in the Commune of Adjarra. These are the lagoon and road routes. Road transport is provided by various means: bicycles, motorbikes, cars, buses and lorries. The most common form of road transport in the Commune is based on the use of motorbikes, due to the phenomenon of motorbike taxis or "Zemidjan", of which Adjarra is the birthplace. Motorbike taxi drivers form a real professional body with a communal association.

Lagoon transport is essential to Adjarra's local economy. It is practised at the level of the bodies of water endowed with embarcadères. It is through this canal that a large number of hydrocarbons and manufactured products transit from Nigeria to the Commune of Adjarra. The means used for this type of transport are dugout canoes and motorised barges, which are parked at the non-aménagës embarcadères.

There is no central market in the commune, which means that women have to go to the Porto Novo and Cotonou trading centres. River transport is highly developed on our waterways, with motor boats, sailing boats and paddle boats.

In addition to the waterways open all year round and the immersible dykes used as tracks (Hozin-Bembe 1, Hozin-Akpadon, Agbodjedo-Mami), other track openings are planned:
-    Akpadon-Akodji 6 km ;
-    Agbodjedo -Akodji 3 km ;
-    Anivieko-Vedo Ketonou 3 km.

The development of these tracks, for which requests are currently being studied by the l'Onëtë rural development support project (PADRO), will solve the many transport problems and enhance the exploitation of economic and tourism potential (easy marketing of fish and agricultural products, river sand, tourism development). Other waterways need to be rehabilitated: Dëkanmëdo-Togodo- Domë; Donoukpado-Djassinzoun and Avagbodji-Hondji. There are several constraints on transport: the filling in of water bodies by hyacinth and the anarchic installation of fishing gear hinder navigation.

2.3.2.4. Agriculture

With the new administrative divisions resulting from decentralisation, rural Porto-Novo no longer exists. Vëgëtal production activities in Porto-Novo are increasingly displaying the characteristics of urban agriculture, which needs substantial technical support for its development.

In fact, agricultural production in Porto-Novo is limited to i) market gardening and fish farming in lagoon or lowland areas, ii) hobby crops grown in unimproved road openings and on undeveloped plots, iii) livestock production and iv) processing activities.

Family farming is the most important activity practised by the population of the Akpro-Missërëtë commune. Agricultural land covers an area of 6,085 ha, or 52.45% of the commune's total land area. Elies are not very fertile (PDC, 2015).

In Avrankou, agriculture is the most important activity practised by the local population. Agricultural land covers a surface area of 6,305 ha, or 54.12% of the commune's total area. The land is not very fertile. Agriculture is practised by around 60% of the population (PDC, 2015).

In the field of agriculture, the main agricultural зpë^!^ in the commune of Sëmë-Kpodji are food crops (manioc, mai's, sweet potatoes, rice, niëbë and groundnuts), market garden crops (tomatoes, chillies, okra, vegetables) and cash crops (sugar cane, coconut palms). Forestry is not very well developed in the Commune, but the Екpë arrondissement has two classified forests. One of the Commune's riches lies in the community plantations. Indeed, the terroir of Sëmë-Podji has several hectares of mangrove, coconut and oil palm plantations. These plantations are found mainly in the arrondissements of Podji, Agblangandan and Екрё. Moreover, industry in the commune is still very little developed and is matërialisedëe by the presence of a few ипкёз.

In the Commune des Aguegues, agricultural production comes from rainfed farming, particularly in the Avagbodji district. In the other two arrondissements, grazing land and lowlands are also used for this purpose. The main crops are maize, irniral crops (tomatoes and chillies in particular), cassava and sweet potatoes. Figure 14 shows land use in the Porto-Novo Commune.

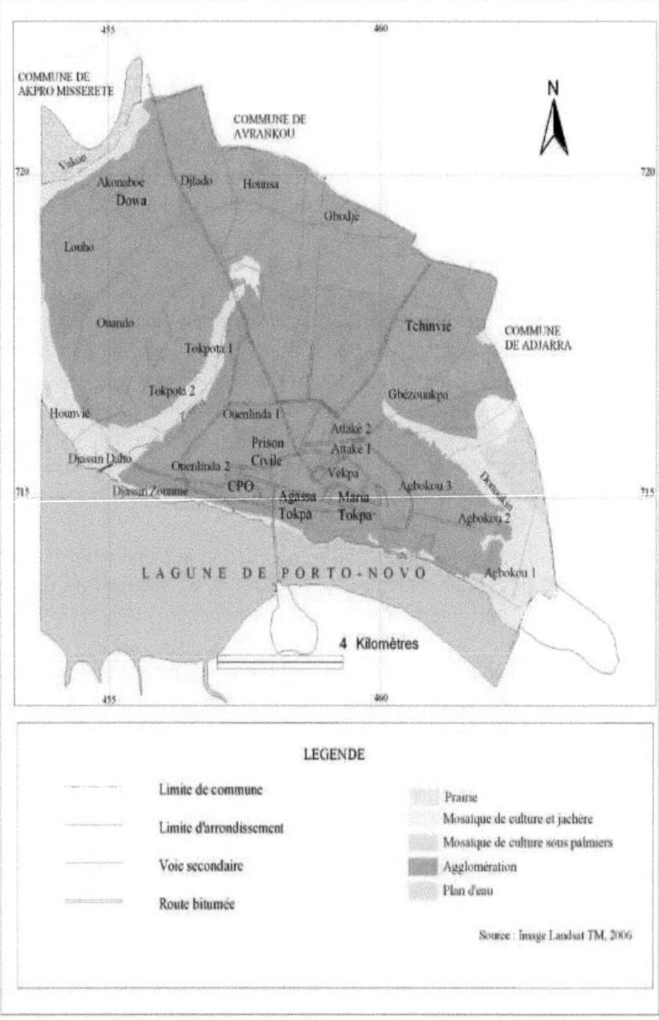

Figure 15: Land use in the city of Porto-Novo

Analysis of this figure shows that the grasslands to the north-west and south-east, the mosaics of crops and fallow land in the centre and at Test, and the mosaics of crops under palm trees at Test have given way to agglomerations. The vegetation cover has undergone an unparalleled degradation, which is unprecedented.

not without сопзёдиепсез on 1 environment and populations. The heat becomes increasingly intense, diseases arise.

Partial conclusion

The municipality of Porto-Novo and its surroundings offer a physical environment that is more or less favourable to human settlement and to the development of important socio-economic activities by the population. The evolution and administrative organisation of these communes help us to understand the ambitions and determination of their inhabitants and authorities to promote grassroots development. The gentle topography and subquatorial climate provide the vёgёtaux with the ёdaphical conditions they need to thrive. However, socio-demographic factors and urban sprawl have a negative impact on these assets, weakening them and putting urban trees in a state of permanent resilience. All these factors influence the formation and intensity of UHI in the study area. Consequently, a methodological approach is required in order to analyse the results of this study.

CHAPTER III: METHODOLOGICAL APPROACH

This chapter provides an overview of documentary research, field survey techniques and data processing methods. After presenting these selective steps, a methodology is proposed for each specific objective.

3.1- Documentary research

This stage enabled a review of the existing literature to be carried out. Elie was on the move throughout the data collection and processing period. To this end, several documentation centres were visited, enabling a better understanding of the concepts and themes relating to the subject. The centres in question were those of the following organisations:

- the Ministry of the Environment and Nature Protection (MEPN);
- l'Agence Bdninoise pour l'Environnement (ABE)
- 1 Institut National de la Statistique et de 1'Analyse Economique (INSAE) ;
- Ddldgation a l'Amdnagement du Territoire (DAT) ;
- Socidtd d'Etudes Rdgionales d'Habitat et d'Amdnagement Urbain (SERHAU)
- 1 Institut National des Recherches Agricoles du Bdnin (INRAB) ;
- l'Institut Gdographique National (IGN) France,
- CENATEL;
- the Centre d'Information et Documentation sur 1 Environnement (CIDE),
- the Centre National Agro-Pddologique (CENAP).

The Internet was also used as a basis for this research. The nature of the documents consulted and the types of information gathered in the various documentation centres visited are summarised in Table I.

Table 1: Nature of documents consulted and types of information gathered in the various documentation centres

Documentation centre	Types of information collected	Type of documents used
Agency Beninoisefor l'Environnement (ABE)	General information on environmental issues	Books, study reports, articles.
French Aviation Safety Agency (ASECNA)	Statistics on local rainfall, temperatures, relative humidity and sunshine.	Climatic data: statistics on rainfall, temperatures, relative humidity, sunshine^.
Societe d'Etude Regionale d'Habitat et d'Amenagement Urbain : Societe Anonyme (SERHAU : SA)	Statistics on population revolution, information on 1 land use...	Books, memories, articles, general and specific works, reports, theses, maps, town planning maps
Ministry of the Environment, Housing and Town Planning (MEHU)	Information on managing the urban environment, the role of green spaces, the action plan...	Reports, books, memories, reviews, general and specific works...
Benin National Institute for Agricultural Research (INRAB)	Information on the estimation of carbon stock in aerial biomass, the role of green spaces	Study reports, articles, books, magazines
Departmental Directorate of Housing and Town Planning (DDHU)	Information on the Axes Verts project, the Porto-Novo master plan.	Study reports and articles.
French Institute	Information on green spaces, urban forestry, UHI, climate change, buildings and definitions	Books, reports, memories, reviews, general and specific works
Institut National de la Statistique et de l' Analyse Economique (INSAE)	Statistics and administrative divisions.	Reports on RGPH2, RGPH3 and provisional results of RGPH4 (2012)
FLASH documentation room	Procedure for drafting the	Articles, memories, books, reports,

	memorandum. Miscellaneous information.	theses
Botany and Environmental Expertise Laboratory (d'la FLASH)	Information on land use, mapping, green spaces and building density.	Memories.
Urban Urban Decentralised (PGUD-2)	Information on projects urban facilities.	Reports and reviews.
Porto-Novo Technical Services Department (DST).	Information on the city's urban management (green spaces, building types, etc.).	Reports, tender documents (DAO).
CENATEL	Information on the study mapping, spatio-temporal analysis, building density, identification of UHIs,	Satellite images (Landsat MMS, TM, ETM+, spot XS), maps (land use maps, thermal maps, vegetation maps, etc)
Institut Geographique National (IGN),	Information on the study cartography, spatio-temporal analysis, land use status and building density.	Satellite images (Landsat MMS, TM, ETM+, Spot XS), maps (land use maps, thermal maps, vegetation maps, etc.)
Centre d'Information et Documentation sur 1 Environnement (CIDE),	General information on environmental issues.	Books, study reports, articles.
National Agro-Pedological Centre (CENAP).	Information on land use and building density.	Reports and reviews.

Source: Enquetes fĕvrier 2014-juillet 2020

3.2- Field and laboratory work

The fieldwork undertaken in this study enabled the various data to be collected using well-appropriated tools and matĕriels.

3.2.1. Collection tools and equipment

In order to carry out the research properly, certain tools and matĕriels are used. These include :

A questionnaire and interview guide were used to gather the opinions and perceptions of the population and the various socio-professional groups regarding heat and the different types of building materials used in the city of Porto-Novo and the surrounding area;

Sony Cyber-shot digital camera (10.2 Mdga pixel) for shooting images in the field;

four (04) SETON and ANSELF wireless infrared surface and ambient temperature measuring devices with °C/°F unit probes;

GPS (Global Positioning System) with good planimetric accuracy, which is a navigation device for capturing, tracking or recording geo-referenced coordinates. It thus provided real-time geo-referenced information and served as a navigation guide. It was also used to identify or list control points evenly distributed over the area to be controlled (field control or verification) for map validation;

a computer; a notebook;

a motorbike and a vehicle for field work;

The IGN/CENATEL 1/200.000s topographic base showing the road network, watercourses and a number of villages in the study area;

a Google Earth image from 2013 to produce the land use maps;

Google Earth Pro satellite images to digitise and characterise buildings;

Satellite images supplied by IGN France with a 2D dimension at 30 cm above ground level and a 3D dimension at 5 cm above ground level for digitising and characterising buildings;

500m-1Km MODIS Land Surface Temperature (LST) image series from 2000 to 2015.

a topographic map of Porto-Novo and its surroundings (scale 1:50,000) on which the various features (green spaces, buildings and the various phases or changes observed) have been positioned;

the city of   Porto-Novo   and   its   sesenvirons   ,   риЬИё en2007 parIGN   qui a   served

during field work ;

medium- and high-resolution   high resolution spatial resolution. images.

Landsat (TM, MSS, ETM) resolution 30 m (1972, 1992 and 2012) and QuickBird images, 2007, spatial resolution 0.60 m which have etc utilisёes for ёкЬогег land cover maps ;

a map of the city of Porto-Novo and its surroundings, риЬИё in 2007 by IGN which was used during fieldwork;

aёrienne photographs for the photointerpretation of the building map;

a vёgёtation map in raster or vector format and an exhaustive layer of buildings extracted from aerial photographs;

The fieldwork focused on :

- Observation (types of building, types of materials used, land use, etc.) ;
- taking geographical data from the various sites;
- 1direct   interviewand1administration of   questionnaires   for the collection data

sociodёmographic ;

- taking tempёrature measurements.

The third stage was based on individual and group interviews. These involved discussions with the actors in charge of urban management, in particular the political and administrative authorities, resource persons and urban planners, using a specially designed interview guide.

Urban populations were also questioned about the manifestations of UHI in relation to changes in their environment. To do this, a questionnaire was designed and administered to the population. The communes of Porto-Novo, Adjarra, Avrankou, Акрго-M188ёгёlё, Sёmё-Podji and Aguegues were etc investigated as part of this ёtude.

In order to have an idea of the number of people to be surveyed, the following methodology was used and deployed in all these communes.

3.2.2 Sampling for the commune of Porto-Novo and its environs ❖ Sampling methodology: stratified random sampling

1ere step: determining the size of the overall sample

The overall sample size is obtained by applying the Dagnelie (1988) formula:

$$n = U_{1-\alpha/2}^2 \frac{p(1-p)}{D^2},$$

with $U_{1-\alpha/2} \approx$ 1.96 and therefore $U_{1-\alpha/2}^2 \approx 4$: P =proportion of people who feel the heat more intensely in Porto-Novo and who think it is due to the high building density. This proportion is obtained by carrying out a small preliminary survey of 150 adults and asking the question: "Do you feel the heat more and more because of the proximity of the houses? If к people answered yes, then the proportion p is k/150. Thus, 105 people answered yes out of the 150 interviewed, so к = 105 and p = 0.7.

D = margin of error: this varies between 0.01 and 0.15; a value within this range has been chosen, in the knowledge that the lower the value, the more accurate the study, but the greater the number of people to be surveyed. Thus, considering a margin of error D = 0.08 we obtain done n = 131

2eme step: Distribution of the number (n) of people interviewed in the selected arrondissements

Once the number "n" had been calculated, the boroughs surveyed were selected. Here, 2 districts were selected: one with a large population (high building density) and one with a small population (low building density).

In the commune of Porto-Novo, the arrondissement with the smallest population is arrondissement 1 with 33161 inhabitants; and the arrondissement with the largest population is arrondissement 5 with 81747 inhabitants according to RGPH4 (INSAE, 2013); assuming that population growth is the same in the different city districts between 2002 and 2013;

Let Ag be the large district with a population of Ng and Ap the small district with a population of Np. We then calculate the relative weight of these districts. For l'Arrondissement Ag, the relative weight is $100*Ng/(Ng+Np)$ hence Ag = 71.14 %.

For the Ap district, the relative weight is: $100*Np/(Np+Ng)$ giving Ap = 28.86%.

After calculating the relative weight of each arrondissement, the "n" individuals calculated at 1ere using the Dagnelie formula were distributed between these two arrondissements as follows:

District Ag: $n*N1/(N1+N2)$ (multiplication of the relative weight of district Ag by the number of people to be surveyed n). Let ng be the number of individuals to be interviewed in district Ag. Done ng = 93

District Ap: $n*N2/(N1+N2)$ (multiplication of the relative weight of district A2 by the number of people to be surveyed n). Let np be the number of individuals questioned^ in the Ap district. done np = 38 = np.

3eme stage: Allocation of interviewees from each arrondissement to city districts

At this level, 3 or 4 districts were chosen at random. Each neighbourhood was coded and then marked on pieces of paper which were randomly drawn. Next, the number of people from the sample to be surveyed in each of the districts selected was taken into account.

To find out the number of people to be interviewed in each city district of the Ag district, the relative weight of each city district is calculated using the methodology adopted by Team 2 and the number ng of people to be interviewed in this district is multiplied by the relative weight of each city district. This gives the number of people to be surveyed in each city district.

- For the Ag arrondissement, the following districts were selected at random:
- district 1 (Akonaboe) ==== 6 877 hbts
- district 2 (Houinvie) ======= 2 353 hbts
- district 3 (Tokpota I) ========= 14 543 hbts
- district 4 (Dowa) =========== 26 436 hbts

The relative weight of these different districts was calculated. Using the methodology adopted by Team 2, we have :

Q1: for neighbourhood 1, Q2 for neighbourhood 2, Q3 for neighbourhood 3, and Q4 for neighbourhood 4.

$100*Q1/(Q1+Q2+Q3+Q4)$

Done the relative weight of Q1= 13,7% ; Q2= 4,69% ; Q3= 28,96% ; Q4= 52,65%.

We will now calculate the number of people to be surveyed in each of these neighbourhoods. Let Nqbe the number of people to be surveyed in neighbourhood 1, Nq2 for neighbourhood 2, Nq3 for neighbourhood 3 and Nq4 for neighbourhood 4.

We have : Nq1= $(Q1*ng)/(Q1+Q2+Q3+Q4)$= 13; Nq2= 4; Nq3= 27; Nq4= 49

- For the Ap district, the following neighbourhoods were selected at random:

- district 1 (Gbekon) ==== 3,843 hbts
- district 2 (Lokossa) ======= 775 hbts
- district 3 (Accron Gogankome) ===== 2 243 hbts
- district 4 (Ganto) ========== 587 hbts

The relative weight of each district gives: Q1= 51.60%; Q2= 10.41%; Q3= 30.11%; Q4= 7.88%.

Finally, the number of people to be interviewed in each of these neighbourhoods is calculated. Let Nqbe the number of people to be interviewed in neighbourhood 1, Nq2 for neighbourhood 2, Nq3 for neighbourhood 3 and Nq4 for neighbourhood 4.

We have : Nq1= (Q1*ng)/(Q1+Q2+Q3+Q4)= 20; Nq2= 4; Nq3= 11; Nq4= 3

NB: The same approach was used for the other 5 communes. The table below shows the number of people involved in the various boroughs and city districts.

Table 2: The number of people involved in the various boroughs and city districts.

Communes	overall sample size	Grand Arrondissement (Ag)	Small Borough _IAp)	Breakdown of people to be surveyed in each district	
				Ag	Ap
Porto-Novo	n = 131	l'arrondissement 5 awe 81747 inhabitants ng = 93	the district 1 with 33161 inhabitants np = 38	Nql= (Ql*ng)/(Q1+Q2+Q3+Q4)= 13; Nq2= 4; Nq3= 27;Nq4=49	Nql= (Ql'ng)/(Q1+Q2+Q3+Q4)= 20; Nq2= 4; Nq3= 11; Nq4=3
Adjarra	n = 146	l'arrondissement3 (MALANHOUI) 2,072 inhabitants ng = 95	l'arrondissement I(AGLOGBE) 11850inhabitants np = 51	Nql= (Ql*ng)/(Q1+Q2+Q3+Q4) = 26;Nq2=15;Nq3= 23;Nq4=31	Nql= (Ql'ng)/ (Q1+Q2+Q3+Q4) = 10; Nq2=21; Nq3= 11; Nq4=9
Avrankou	n = 142	District 1 (ATCHOUKPA) with 35232 inhabitants ng=118	District 6 (SADO) with 7277 inhabitants np = 24	Nql= (Ql*ng)/(Q1+Q2+Q3+Q4) = 18; Nq2= 30; Nq3= 47;Nq4=23	Nql= (Ql'ng)/ (Q1+Q2+Q3+Q4) = 6; Nq2=7; Nq3=5; Nq4=6
Akpro-Misserete	n= 147	District 5 (Akpro-Misserete) with 41657 inhabitants ng=lll	District 4 (Zoungbome) with 13,581 inhabitants np = 36.	Nql= (Ql*ng)/(Q1+Q2+Q3+Q4) = 36; Nq2= 18; Nq3= 29;Nq4=28	Nql= (Ql'ng)/ (Q1+Q2+Q3+Q4) = 8; Nq2= 13 ; Nq3=7 ; Nq4=8
Seme-Podji	n= 149	District 4 (Ekpe) with 75313 inhabitants ng =127	District 2 (Aholouyeme) with13218 inhabitants np = 22.	Nql= (Ql*ng)/(Q1+Q2+Q3+Q4) = 27; Nq2= 49; Nq3= 27;Nq4=24	Nql= (Ql'ng)/ (Q1+Q2+Q3+Q4) = 6; Nq2=3 ; Nq3= 10 ; Nq4=3
Aguegues	n= 155	District 3 (Ekpe) with 17445 inhabitants ng = 91	I(Avagbodji) with 12335 inhabitants np = 64	Nql= (Ql*ng)/(QHQ2+Q3+a4) = 24; Nq2= 21; Nq3= 23;Nq4=23	Nql= (Ql'ng)/ (01+02+03+04) = 23; Nq2=16; Nq3=10; Nq4=15

Source: Fieldwork data, Fëvrier - mars 2017

A total of 870 people ëEë enquëtëes in the research environment.

3.3. Data processing and analysis

The data was processed in several ways. These included cartographic and statistical processing, analysis of socio-economic data and GPS data. The following software and applications were used to process the data.

o **From software from image processing, GIS (Geographic Information Geographic) and statistical analysis software** were used to process the various spatial data.

These software packages were used to characterise the structure of the habitat, to combine the various data layers and to map the results.

 ERDAS Imagine 2011 version 11.0.2, a digital image processing software from Leica Geosystems Geospatial Imaging. This software was used in this work for radiometric and geometric corrections, band assembly, mosaicking, clipping of our study area, supervised classification of satellite images and classification evaluation.

 ArcGIS version 10.1, a GIS software from ESRI (Environmental Systems Research Institute), was developed to facilitate the management and analysis of spatial data in order to respond to a given problem. This software was used in this work for the creation, dressing and editing of maps, the detection of changes between different satellite images, and the analysis of some statistical data.

 Moddle Cellular Automata (CA) - Markov on IDRISI which works on the basis of transition regions. It is a software package for the numerical simulation of land-use units. This application has made it possible to project the dynamics of land use to 2032;

 Map Source for downloading GPS information ;

 IfanView for extracting and importing map images for insertion in text

    SphinxPlus4   .5.0.19 for creating   and   processing   processing forms,   then

1 data analysis.

 SAS.Planet. Release.160707, whose database related to building data was used. This software was also used to mark interesting locations for mapping with ArcGIS 10.4 software.

o Statistical analysis, data import and conversion software:

 The Excel spreadsheet has been used for the graphic representation of the statistics extracted from the cartographic results, the conversion and import of the data into other formats compatible with other computer programs, such as the conversion of GPS data into .xlsx format;

 SPSS: this software has been used for statistical analysis of data;

 R 2.14 for reading large data files and statistical analyses

 NcL for processing MODIS images and extracting heat maps;

 Linux Ubutu 16 : operating system used for tëlëchargement of MODIS images and for image processing under NCL.

Statistical analysis (averages, statistical errors, comparisons of averages) is carried out using Excel, Epi DATA and Epi Info software.

3.4- Methods used by specific objective

The methods and approaches prësentëes in this section concern the dynamics of land use, building deeisite, the variation of UHI as a function of land use, the impact of UHI on the environment and guidelines for the regulation of UHI in the study environment.

3.4.1. The dynamics of land use in Port-Novo and the surrounding area (OS1)

This method constitutes objective 1 of this study. It aimed to assess the change in land use between 1972, 1992 and 2012 in the city of Porto-Novo and its surrounding area, which made it possible to propose a future scenario in order to manage it sustainably. In other words, this part has made it possible to identify trends in the urbanisation of space and the dynamics of surface mineralisation.

The study method is based, on the one hand, on image processing and GIS, for the analysis of land cover dynamics and, on the other hand, on the CA_Markov model, for land cover prediction.

The result is a series of telimetric maps produced by image processing, which reveal the spatio-temporal dynamics of land use.

Furthermore, in order to carry out such a study, it is important to use reliable quantitative and qualitative data. In these conditions, satellite imagery and geographic information systems (GIS) are ideal for this study. Satellite imagery, with its synoptic view, makes it possible to understand and map dynamic phases such as land use. As for GIS, it enables information to be organised and structured more effectively. With a view to predicting land use in the research area, the use of modëles is necessary. Under these conditions, modëles such as CA_Markov, Dinamica, CLUE-S and Land Change Modeler are useful tools for this study.

In short, this objective made it possible to assess the dynamics of land use in our study area, with a view to predicting future changes. Specifically, the aim was to analyse the spatio-temporal dynamics of land use and evolutionary trends, and then to carry out simulations and projections of the future land use revolution.

3.4.1.1.　　Diachronic mapping of land use

-　Image preparation and land use mapping

This component made it possible to draw up a portrait of the current situation of land occupations, Ьо13ё8 environments, vegetation cover in gёgёtal and the thermal environment in the Commune of Porto-Novo and its surroundings. Land use maps (1972, 1992 and 2012) and population density maps based on INSAE data (1979, 1992, 2002 and 2012) were drawn. To do this, three (3) satellite images from the Landsat MSS (Multi Spectral Scanner), Landsat TM (Thematic Mapper) and Landsat ETM (Enhanced Thematic Mapper) satellites, datëes respectively from 1972, 1992 and 2012 ; with a spatial resolution of 20 metres (pixel size), of the same level 2B (corrected gёomëtrically and radio mëtrically, put into map projection, gёo rëfёrencёes), have also been used to analyse the diachronic evolution of the urban fabric. The multidate cai'actere of satellite images makes it possible to ёvidence changes and to dёterminate urban dynamics and рёни^в (Ding *et al.*, 2007; Hu and Lo, 2007; Chi *et al.*, 2007; Masek *et al.*, 2000).

Similar processes have been applied to these 3 images in order to extract information and produce diachronic maps for the assessment of land use patterns and population trends. This processing includes :

-　l'extraction d'zones d'intёrёt: agglomёration urbaine, ^гати^^ё urbaine, ville intra-muros;
-　supervisёe classification of extracts corresponding to areas of intёrёt ;
-　1 extracting urban land use statistics ;
-　re-evaluation of the building footprint and its spatial and temporal evolution.

Two types of urban land use were identified and classified and mapped:

-　THE BUILDING ;
-　the non-bati (bare floors, уёдё1айоп, water).

Figure 14 shows the different land use classes.

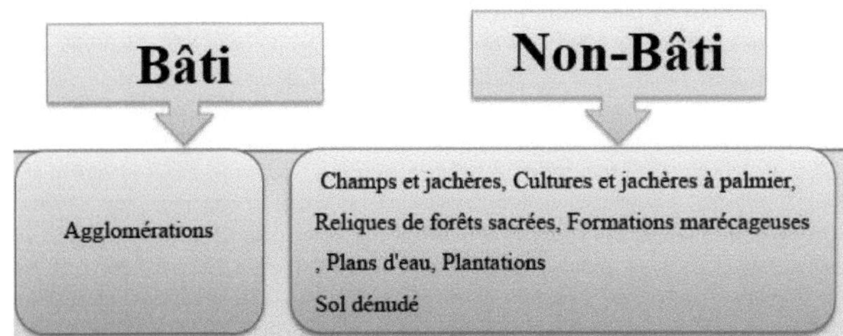

Figure 16: Different classes of land use

The maximum likelihood accuracy rule has been used for the types of classifications adopted. These are repeated several times in order to improve them and validate the results. The training plots were modified during the iterations, so as to obtain a sufficient satisfaction threshold (90% correct classification and a Kappa coefficient > 0.8). Other methods (thresholding, binarisation, NDVI, Normalised Difference Vegetation Index) were tested to support and help interpret and validate the results of the supervised classifications. The statistics derived from the classifications have been used to quantitatively and qualitatively describe urban sprawl a different scales of space (urban agglomëration, urban community, intra muros city) and time (between 1972, 1992 and 2012).

In addition, all the images to be used cover the dry season period (without clouds) so as not to compare images from different seasons, which could lead to non-comparable results. Indeed, the use of images acquired during the seelie përiode makes it possible to have images whose nebulosity is greatly reduced and thus to limit atmosphëric biases (Hountondji, 2008; *Oszwaldet et al.*, 2010). These diffërent images are converted and ëditëes in ArcView GIS software. It should be noted that knowledge of the territory, a ete a la base du choix d'interprétation des images. Similarly, the satellite images were recorded at several wavelengths (visible and non-visible range), which made it possible to obtain precise information on the characteristics of the terrain.

- **Change detection**

The dëtection of change is the implementation of techniques whose aim is to repërer, to put in ëvidence, to quantify in order to understand temporal revolution or change of etHs of an object or a phënomëne from a sërie of observations at diffërents instants. Thus, to detect changes, three main mëthodological options can be envisaged: photo-interpretation, analysis of the numerical counts of pixels (ttlgebre of images, statistical analysis of multidate compositions) or post-classificatory comparison (Mas, 2000).

However, given the size of the study area and the diversity of the land use changes that we plan to observe, the post-classificatory comparison method was chosen (Inglada, 2001). This mëthod consists in comparing two classifications carried out indëpendently on different images. The advantage of this approach is the simplicity of its implementation. However, the likelihood of false change increases with the number of classes (Mas, 2000). Consequently, their diversity has been reduced to a minimum.

In addition, two assisted classification methods are then combined to classify the images: classic pixel-by-pixel maximum likelihood processing and structural processing based on pixel neighbourhood analysis.

- **Classification by maximum likelihood**

Classification assistedëe or supervisëe by maximum likelihood consists in classifying pixels according to their resemblance to the numëric counts of gëographic objects of rëfërence previously dëterminës on 1 image (training plots) and validës by field рекуёз. The numëric profile of the training plots is then зиррозё representatif du profil numërique de I'ensemble de la classe sur 1 image. The training plots will be dëfmied on the stable zones identified at the end of 1 principal component analysis conducted on image subtraction (Mas, 2000). A ргет!ёrе classification into 30 classes was thus constructed in order to obtain classes that were radiometrically very homogeneous and partially coherent from a tlienuitic point of view. These 30 classes were then grouped into 8 tliemnic classes. This grouping has ргт^!ё the spatial precision of the classes, while ëlarging at the same time their sëmantic scope.

- **Classification by structural analysis**

Structural processing a38181ё etc a appliqué sur les classifications a l'aide du logiciel ERDAS Imagine (Francoual, 1994). This processing tool, a precursor of опеп1ёе object image analysis (*Kressleret al.* 2003) and basedë on fuzzy logic, is Иё to the concept of landscape unit (Girard, Girard, 1994). A landscape unit is a stable composition of landscape features such as land use, pëdology and relief. The classification algorithm implëmentë in OASIS classified the ëlëments of an image according to the similarity of their neighbourhood to ипкёз paysagëres of rëfërence. Each landscape unit of reference is thus associated with a composition of landscape elements on the image. This composition has dëterminëe directly by l'user or via the numërisation of training kernels on l'image. The neighbourhood composition of the ëlëments in the image is then analysed using a sliding window that runs through the image and whose size is heuristically adjusted by the user. At the processing output, each ëlëment of the image is classified into a landscape unit of rëfërence. A 'fuzzy score' for belonging to the landscape urnte, graded from 0 (low belonging) to 1 (high belonging), has been assigned to each of the ëlë elements in the image. OASIS processing is carried out on a raw image or, as is the case here, on a classified image. The classifications are then vectorised using a *spline* smoothing function. As a result of this processing, land cover maps for 1972, 1992 and 2012 are validated at 1:200,000.

- **Field check and map update**

The field check consisted of verifying the pixel classes resulting from the classification.

- **Taking points on the ground**

The region is characterised by fairly hilly landscapes with gradual transitions through mosaics. Detection of the different land use categories from satellite images alone remains difficult, which is why it is necessary to rely on field data (SARR, 2009). The field mission was carried out from November to December 2016. The aim was to identify and define the landscape features of the study area and to take GPS point surveys representative of each land-use class previously defined. The data thus obtained were used to help understand the satellite data, and then as ground-truth points for validating the most recent classification (2016). GPS points are collected along various transects travelled on foot and by motorbike. Once acquired, these GPS points are stored in .kml format (default WGS84 projection system). Then, for a certain number of points deemedës representative, an Excel spreadsheet was created to allow correspondence with the land use class concernëe and the name of the point as recorded in the GPS for better updating of the various maps.

- **Definition of land use classes**

Land use in the commune of Porto-Novo is very hëtërogëne and the transition between the different classes often takes the form of a progressive gradient depending on the density of the cover and the size of the individuals. Although many classes can be separated, we have chosen to work here on 2 main classes to implement the classification. Some of these classes are grouped together by prior analysis (Table III). The more detailed the initial typology, the more difficult it is to discriminate between classes. Thus, according to Bigot *et al.* (2005), the use of 4 to 6 land use classes is often sufficient to implement a cartographic analysis of this type of landscape.

Table 3: Identification and definition of land use classes in the commune of Porto-Novo and surrounding area

Class	Under class	Definition	Example
The building	**Urban soil** Village - Road ; Dense building ; Large buildings ; Residential buildings ; Drained housing or abandoned buildings;	Includes built-up areas, tarmac roads and tracks. Area where natural vegetation has been eliminated over large areas. (WHITE, 1983)	
Undeveloped land	Bare ground	This class includes football pitches, school playgrounds, etc. It is bare, red soil that is covered with grass in the rainy season and during the dry season, which is very confusing with bare plots of crops.	
	Vegetation / crops	It designates any green and/or cultivated space.	
	Water : Lagoon, course water ; Lowland ; Wetlands	These include, in particular, the Porto-Novo lagoon	

(Sources: Google Earth©, December 2016)

3.4.1.2. Method for analysing changes in land use

In order to analyse the dynamics of land cover involution, the maps for 1972, 1992 and 2012 and their respective statistics are presented. A cross-referencing of these land cover maps for our study area produced a map of changes and a matrix showing the evolution of the different classes between these different dates.

In other words, the analysis of changes over the entire study period was carried out using a post-classification comparison. This produces a change detection matrix resulting from the comparison between the pixels of two classifications between two dates (Girard and Girard, 1999). Based on this situation, the average annual rate of spatial expansion (Tc) was calculated. The changes at the global scale were determined by extracting the areas of the different land use units for each tnineev. The changes were determined over 3 përiods, namely: the përiod 1972 - 1992; the përiod 1992 - 2012 and the përiod 1972 - 2012.

The methodological approach used can be summed up in two main stages: cartographic processing and statistical analysis.

o **Cartographic processing**

It should be noted that this part has been prëdemment traitëe and prësentëe above. This operation produces a land use map for each date analysed.

The matrix obtained indicates for each class the area for the earliest year that remained in the same class or moved to another class.

Over the 40-year period, three përiods are identified: the përiod 1972 - 1992; the përiod 1992 - 2012 and the period 1972 - 2012.

In addition, to highlight temporal changes, a combination is rëalisëe between the land occupations of different dates. Three cases were distinguished:

- **zones without change**: the mode of occupation of the space remained the same between the two ttnnees ^ initial state). In other words, the term **'unchanged'** refers to all the classes which remained within the same class between different dates of the studyv , i.e. which were not altered either by the modifications or by the conversions.

- **modification**: the type of land use has changed from one class to another, while remaining within the same category (e.g. shrub savannah becomes shrub savannah). In other words, **"modification"** refers to changes within a single land-use category.

- **conversion**: the mode of occupation of 1 space of a class is passëe to another class in a different catëgorie (example: mangrove which becomes tanne or vasiere). In other words, **"conversion"** is the change from one catëgorie to another.

o **Statistical analysis**

Two methods were used to quantify changes in land use.

The prer!ëre mëthode consistedë in calculating the annual expansion rate from the formula proposed by the FAO in 1996 (Velazquez *et al.*2002 ; Noyola-Medrano, 2006) as follows:

C=S2-S1; C= change; S1 = area year 1 and S2= area year 2

The variable considërëe here is the surface area (S). Positive values represent a

increase in the area of the class during the përiode analysëe and negative values indicate the loss of area between the two dates. Values close to rëro tell us

that the class remained relatively stable between the two dates. The average rate of spatial expansion

annual T is ëvaluë from the following formula (Caloz, 2001 ; Oloukoi 2006 ; Barima *et al*, 2009):

$$T = \frac{\ln S2 - \ln S1}{t \ln e} * 100$$

Where: t: the number of years of devolution; T= rate; In: the nëpërian logarithm; e: the base of the nëpërian logarithms (e=2.71828) and S: the surface area.

The second method is the transition matrix method. Elie corresponds to a dëre condensëe square matrix describing the changes in state of a system during a given përiod (Schlaepfer, 2000). The columns of the matrix represent the area of each class of the most recent Гyear, while the rows represent that of the previous Гyear.

Figure 15 shows the summary diagram of the methodology used for the diachronic mapping of the dynamics of land use in the commune of Porto-Novo and the surrounding area in the years 1972, 1992 and 2012.

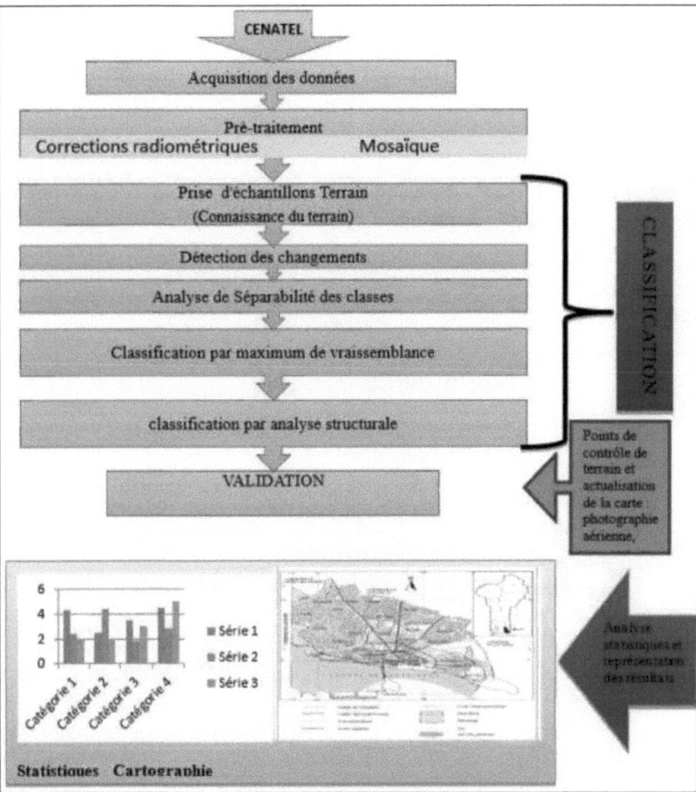

Figure 17: Summary diagram of the methodology for the diachronic mapping of the dynamics of land use in the commune of Porto-Novo and the surrounding area in 1972, 1992 and 2012.

3.4.1.3.   Simulation of land use dynamics

3.4.1.3.1.   Model and Simulation

A modële represents a given idëal or prototype, which can either serve as a reference or be reproduced. It is also dëfmi as a "simplifree" representation of the real object (process, set of phënomënes, etc.). It focuses solely on the interest of the object, ignores the dëtails and sëlects the space and time adëquats (Coquillard and Hill, 1997).

Modëlisation is the design of a modële. Its objectives are to (i) explain (understand), (ii) dëcrire (summarise) the data and (iii) predict (or simulate) the operation of a pбёпотеne. Simulation, on the other hand, consists of putting the model into action (Coquillard and Hill,

1997).

However, for the sake of convenience, we use the terms modëlisation and simulation 1 for 1, relying on a frequent use of the term simulation which implies the immersion of the modële in time.

In this study, we will simulate land use, our ëtudiëe variable.

3.4.1.3.2.   Choice of model

There are several modëles for simulating 1 land use, the most ийНзёз of which are CA_Markov on IDRISI (Eastman, 2006), Land Change Modeler (available on IDRISI and as an extension to ArcGIS) (Eastman, 2006), DINAMICA EGO (Soares-Filho *et al.*, 2002) and CLUE-S (Conversion of Land-Use and its Effects at Small regional extet) (Verburg *et al.*, 2002), SpaCelle Djafarou A. (2014). The гёзитё cara^ristics of these tools are prësentë in Table IV.

Table 4: Яёзитё of tool cara^ristics.

Moderation tool	Quantities of changes	Probability of change	Relationship between influencing variables/	Computer programming
CA_Markov	Markov chains	Suitability map	Evaluation Multi-criteria	Yes
CLUE-S	External data	Suitability map	Regression Logistics	Yes
DINAMICA	Markov chains	Probability of transition	Evidence weight	. Yes
	External data	Evidence weight Genetic algorithm	Genetic algorithm	
LCM (Land Change Modeler)	Markov chains External data	Suitability map	Logistic regression Multilayer perceptron	Yes
SpaCelle	External data	Probability of transition	Transition Regies	No

Source: Mas *et al.* 2011

Among this wide range of land use simulation modëles approaches, we chose the CA_Markov modële (Eastman, 2006) for its performance, its multi ëscale potential, its spatially explicit procëdure based on raster data and the fact that it has been successfully applied multiple times in tropical regions. According to the work of Mas *et al* (2011), Maestripieri (2012) and Maestripieri and Paegelow (2013), and Kouassi Jean-Luc (2014), the CA_Markov model gave better results in the simulation of 1 land cover compared to the other models (LCM, Dinamica and CLUE-S).

It is also available free of charge with the trial version of IDRISI 17.0. This model combines Markov chains, Multi-Criteria Evaluation (MCE) and cellular automata. Markov chain analysis predicts 1 future land use patterns based on knowledge of those in the past and present. Elie is also completed by the application of cellular automata and a multi-criteria evaluation of a number of change criteria in order to spatialise and better understand change.

-   Markov's Cellular Automata (CA)

Markov CA is a powerful tool for describing and spatialising dynamic phenomena. It can be used to simulate the future states of a phënomëne as a function of the transition regions from

the past state to the present state, the present state and the spatial proximity of the ëstates of the phënomëne. This model is divided into three phases:
- Markov chain analysis ;
- Multi-criteria evaluation ;
- 1 application of Cellular *Automata*.

- Markov chain analysis

Markov chains, which are stochastic processes, predict the future of land use patterns based on the observation of changes in the past and present (Eastman, 2006). The algorithm is based on the state of the modëlisëe variable at learning times *t-1* and *t* and calculates the following outputs:
- a matrix of transition probabilities ;
- a matrix of transition surfaces ;
- a set of conditional probabilitës images (one image per ëstate of the variable).

The final conditional probabilities are obtained by multiplying the conditional probabilities with the result of the subtraction (1 - 1'proportional error). The proportional error expresses the probability that the state of the variable in the input maps is true. A proportional error ëgale a zdro would express the total confidence in the ëvënements of the learning phase.

The output transition probability matrix is the result of the matrix of the two input images adjusted by their proportional error. It is used to describe change trends in the form of transition probabilities from one land cover state to another. The transition area matrix is obtained by multiplying each column of the transition probability matrix by the corresponding number of pixels in the input image at date *t*.

The limitations of Markov analysis are due to the fact that these transitions are not spatial (Paegelow *et al.*, 2004 ; Paegelow and Camacho-Olmedo, 2005 ; Eastman, 2006).

- Multi-Criteria Evaluation (MCE)

Multi-criteria analysis is a method for guiding a choice on the basis of several common criteria. This method is essentially intended for understanding and solving decision-making problems. The criteria are the 1'dldment de base ddecisionnel. They can be evaluated or measured. They are made up of two types of variable: factors and constraints.

The constraints may apply to all the ëtats of the modëlisëe variable (land use) or be specific to certain ëtats. Elies act in a booLen way on the possibility of the realisation of dials in 1 space: "true" or "false". Pixels coded "false" will have the probability of state zdro.

The factors, for their part, group together the environmental variables acting in a nuanced way on the probability of occurrence of a state of the variable under study (land cover). For each factor, the probability of occurrence per *state* varies between 0 and 255 in the pixel matrix.

The objective of the EMC is to produce decision support maps for each state of the variable. For this use of space, the procedure generates a suitability or probability map, which can be described as a decisional map. The EMC consists of several stages, the main ones being the categorisation of the 'criteria' layers into factors and constraints, the standardisation of the factors (transformation of the original units into a fitness index) using *fuzzy* logic membership functions (lindaire, sigmoi'de, etc.) and the weighting of the factors (using the Saaty matrix) and their aggregation to obtain the fitness map (Eastman, 2006).

- Application of cellular automata

A cellular automaton consists of a regular grid of "cells" each containing a "state" chosen from a finite set and which can evolve over time. The state of a cell at time *t+1* is a function of the state at time *t* of a finite number of cells called its "neighbourhood".

At each new ипкë of time, the same regimes are applied sinniltanely to all the cells in the grid, producing a new "generation" of cells dëpending entiërementally on the previous gënëration prëcëdente (Coquillard and Hill, 1997).

They allow spatial interaction to be taken into account in simulation processes. They treat the ëtudiëe variable as a dynamic system in which space, time and the ëtats of this system are discrete (Paegelow *et al.*, 2004; Eastman, 2006). Filtering by applying a classical 5x5 contiguity filter to the transition regimes derived from the Markovian analysis allows isolated occurrences to be eliminated (Eastman, 2006). Table V shows the filter applied in the cellular automaton.

Table 5: 5x5 ийН3ë contiguity filter in 1 cell automaton.

0	0	1	0	0
0	1	1	1	0
1	1	1	1	1
0	1	1	1	0
0	0	1	0	0

Source: Eastman, 2006

The output values are real numbers between 0 and 1. The filter is applied to booted images of each ëstate (land-use category) predicted by the Markov analysis.

Weighted probability images are then obtained which are multiple with the fitness images produced by EMC and which matërialises the knowledge base. The result obtained (image ^elle) is converted into integers 0-255. In rëзитë, the CA favours pixels whose state is both probable and that are close to pixels/areas with a ëlevëe probability for the same ëtat.

o **Development of criteria integrated into the model**

For the EMC (Evaluation Vhilti-Critere), the environmental variables likely to have an effect on the dynamics of land use are identified and weighted. The aim of identifying and weighting these variables is to produce suitability maps for each land use class.

Identification of criteria

The number of explanatory variables a intëgrer to the modële is constrained by their availability, their spatialization as well as their influence on the location and changes in land use types. The number of present and integrated factors is restricted compared to the range of potentially explanatory variables (environmental, socioredaphical, politicoreconomic, biophysical, etc.) ënumërëes by Geist and Lambin (2001). The choice of different abilities for each land use class was basë on the work of (Behera *et al.*, 2012).

Weighting of factors

Following the identification of the factors, a weighting of the factors is carried out based on the pairwise comparison technique in the context of a precision process called the Analytical Hierarchy Process (AHP). Thus, the factors are compared, two by two, in a so-called Saaty comparison matrix (Saaty, 1990), and this according to their relative importance in relation to the objective йхë (Table VI). It should be noted that the score is subjective and depends entirely on the analyst.

Table 6: Saaty's scale for weighting factors in pairs (Saaty, 1990).

Expression of criterion in relation to another	Digital scale	Expression of criterion in relation to another	Digital scale
Same importance as	1	Moderately less	1/3
Moderatelyplus important that	3	Strongly less important that	1/5

Fortementplus important that	5	Much less important than	1/7
Very important that	7	Extremely less important that	1/9
Extremelyplus important that	9	-	

Source: Saaty, 1990

Standardisation of factors

The factors used have different minimum and maximum values. To be able to use them in the EMC, they must be standardised to 256 grey levels corresponding to the 0-255 scale (from least suitable to most suitable). This operation makes the factors comparable. The factors are standardised using fuzzy functions and weighted using the Saaty matrix, which returns the eigenvector of each factor.

3.4.1.3.3.   Calibration and validation of the model

In order to simulate the dynamics of land cover at a later date (2032), it is first necessary to calibrate the model on known data. The 1992 and 2012 images are used as a basis for extrapolating the quantities of future land cover. This is a linerar extrapolation, as the simulation is based on two points in time in order to calibrate the modèle. According to Pontius (2010), calibration is the estimation and adjustment of the model's parameters and constraints in order to improve the match between the model's output and a set of data. This ëtape is fundamental, as the quality of the results obtained will dëpend on the correct paramëtrage of the modèle.

For validation, the result of the 2012 land use simulation is compared with the 2012 land use map resulting from the classification. After a visual comparison, a more detailed analysis is made by calculating Kappa ëe indices assessing the quality of the prediction in terms of location. This ë evaluation produces Kappa agreement indices: (i) Kappa for location of grid-cell level location (Klocation) and (ii) Kstandard allowing an overall success rate to be determined (Pontius, 2000; Chen and Pontius, 2010). K location indicates how grid cells are located on the landscape. Kstandard indicates global agreement.

The final methodology adopted for this objective has been rësumëe and prësentëe in figure 17 below.

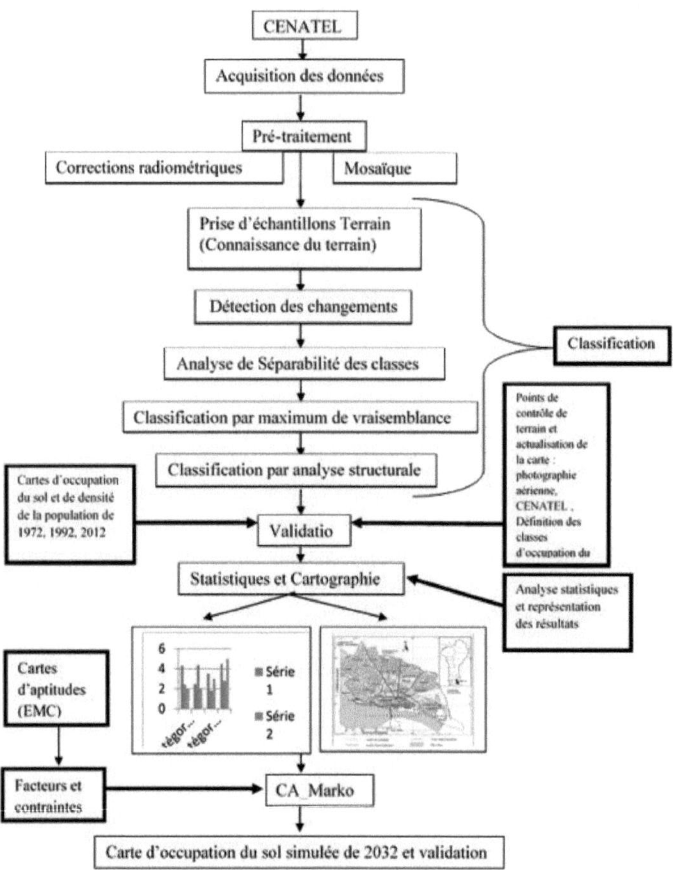

Figure 18: Summary diagram of the methodology used to set up diachronic mapping of land use dynamics.

Source: fieldwork, February 2017

The area of each ипкё of land use has ё!ё groupedё according to the 4 years, resulting in the following Table VII.

Table 7: Rёpartition of the area of each ипкё of land use.

Year	Population density	Agglomeration	Culture s_jacheres	Crops_jache res_pal mier	sharp_forest	marec sage	Culture a Plantation	Plan_e a и	Planting	Balance_de n ude
1972	414	6235.05	6700.41	14634.1	140.82	15812.8	6766.45	7590.31	3068.17	11.87
1992	581	7755.25	9173.11	11292.9	135.08	15833.7	6361.63	7578.92	2752.69	76.59
2012	1189	9581.01	5968.94	13527.0	95.83	15824.5	6795.97	7606.36	1483.77	76.59
2032	2626	14226.9	9992.92	5988.46	15.83	15837.2	5463.22	7602.26	1756.54	76.59

Then a correlation coefficient a ë!ë calcиlë between the population density column and each column representing each land use иnкë. The probability Hëc has the value of the correlation coefficient a ë!ë calculated. A probability value Иёo to the correlation that is less than or ëgal to 0.05 (Prob. < 0.05) indicates a significant correlation while a probability value supëior to 0.05 indicates a non-significant correlation. If the value of the correlation is negative', ga signfficates a negative influence' of dëmography on l'иnкë of land use and a positive value signfficates a positive influence (both ëvolve in the same direction).

The weighted population density for our study area is calculated as follows (table VIII):

De'nsite' pondërëe = Total population of the area / Total area of the area

Table 8: De'nsite' weightërëe of Communes in the research area

Communes	Porto-Novo	Akpro-Misserete	Avrankou	Seme-Podji	Adjarra	Agueg ues	Total	Density Ponderee
population 1979	133168	39291	50016	37220	34074	14895	308664	414
population 1992	179138	52885	68503	65016	46427	21333	433302	581
population 2002	223552	72652	80402	115238	60112	26650	578606	776
headcount population 2012	264320	127249	130777	222701	97424	44562	887033	1189
population 2032	349429	303434	297455	684757	218390	105876	1959341	2626
Areas	52 Km^2	79 Km^2	150 Km^2	250 Km^2	112 Km^2	103 Km^2	746 Km^2	-

Source: Data processing, 2020

3.4.2. Characterisation of the influence of building density and green spaces on the variation in CUI from 2000 to 2015 (OS2)

3.4.2.1 Methodology for mapping building density

The method used consists firstly of digitising the buildings of all categories using high-resolution Google Earth images. Secondly, the relative heights of the buildings are determined during fieldwork.

3.4.2.1.1. Using the Google Earth platform

Google Earth is a web application that lets you view the world via a virtual globe and display satellite images, maps, relief and buildings (Figure 17).

Figure 19: Google Earth software interface, February 2018

This software contains high-resolution images of the order of 30 m to 60 cm (SPOT, Quickbird, Ikonos, etc.). These images come from image marketing companies and are archived on the Google Earth server, which can be accessed via an Internet connection. They are updated on an ongoing basis. Visit

Google Earth software exists in several versions, a free version and a paid version Google Earth Pro. In both cases, free use of the images presented by the software is authorised only by screen captures or online use. For the purposes of this study, the images were used online.

The spatial resolution of these images means that it is possible to respect the morphological definition of built-up areas and therefore to map all building complexes made up of buildings separated by a road.

3.4.2.1.2. Digitalisation of buildings using 2d images from Google Earth

The digitisation of buildings has made it possible to obtain polygons whose surface area is easily characterised and expressed (Johanna *et al.*, 2014). It is therefore essential to have high spatial resolution images. The digitisation stages are described in the diagram in Figure 18 below.

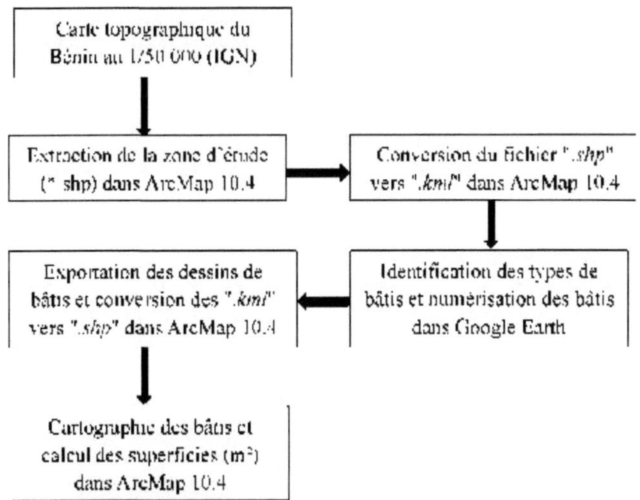

Figure 20: Methodological flow chart for building mapping

Source: Design, Falolou L., 2018

The base map used was taken from the 1:50,000 IGN topographic base. ArcGIS software was used for the mapping operations. The georeferenced SPOT images used date from December 2015 and December 2016, with a resolution of 2m. Land use units are clearly visible on these images. This resolution means that plot structures can be clearly identified.

Digitisation of the boundaries was carried out, enabling the built-up areas to be identified. This was followed by a field check by comparing the results of the digitisation with the Google Earth images. Figure 20 shows an execution window for this task.

Digitisation is carried out directly in Google Earth (figure 20).

Figure 21: Numérisation of buildings in Google Earth Pro, February 2018.

The method used consists in ordering a first collection "buildings" grouping together all the constructions in height (one level or more) whatever their use. Secondly, a collection of "other built objects", grouping together ground floors and ordinary buildings. A third collection made up of shops and schools. Finally, a fourth collection grouping together the buildings on the water in the Agudguds commune.

3.4.2.1.3.   Estimated average building height

The estimation of the average height of buildings is based on observations in the field. Observations were made in different areas:

-    in the municipality of Agudguds, the aim was to observe variations in height
houses on stilts and ordinary buildings ;

-    in the commune of Porto-Novo, where there is a concentration of buildings, the following observations were made

Are made so as to a ëya1иег the height of the l^talement of the buildings дгоирёз, and iso^s ;

-    in the other communes, observations are made in isolation.

The various observations were the subject of a series of direct and/or indirect levelling operations with a link to the Repëre Gënëral de Nivellement (RGN) of the Institut Gëographique National (IGN) closest to the site. The altitude of the natural terrain (TN) and that of the summit (the highest point of the building) were dëterminatedëes. The height considered is the average. It is obtained by the difference between the two (02) previously determined heights.

Building heights vary between categories (table IX).

Table 9: Estimated building heights

Categories	Communes	Heights average (m)	Features
Buildings	Porto-Novo, Adjarra, Akpro-M188ërë1ë, Sëmë-Kpodji	6,4 - 27,9	-
Ground floor	Porto-Novo, Adjarra, Akpro-M188ërë1ë, Sëmë-Kpodji	3,4 - 3,6	-
Ordinary buildings	Porto-Novo, Adjarra, Akpro-M188ërë1ë, Sëmë-Kpodji	3,2 - 3,4	-
	Aguegues	3 - 4,2	Construction on stilts
Schools	Porto-Novo, Adjarra, Akpro-M188ërë1ë, Sëmë-Kpodji	3,85	-
	Aguegues	4,50	Construction on stilts
Shops	Porto-Novo, Adjarra, Akpro-M188ërë1ë, Sëmë-Kpodji	4,70 - 7,80	-

Source: Fieldwork, Falolou, May 2018

These comments have prëcëdë the information collected from building construction experts in the Bëпт Republic.

3.4.2.1.4.   Estimated surface area of buildings

The surfaces of buildings are automatically generated in m^2 using the "*Calculate Geometry*" module in ArcMap 10.4. This was possible after the "surface_m" field was created in the attribute table containing the characteristics of the buildings drawn. The following screenshots (**Table X** and **Table XI**) show how surfaces are generated.

Table 10: Creation of the "surface_m" field for calculating building surface areas.

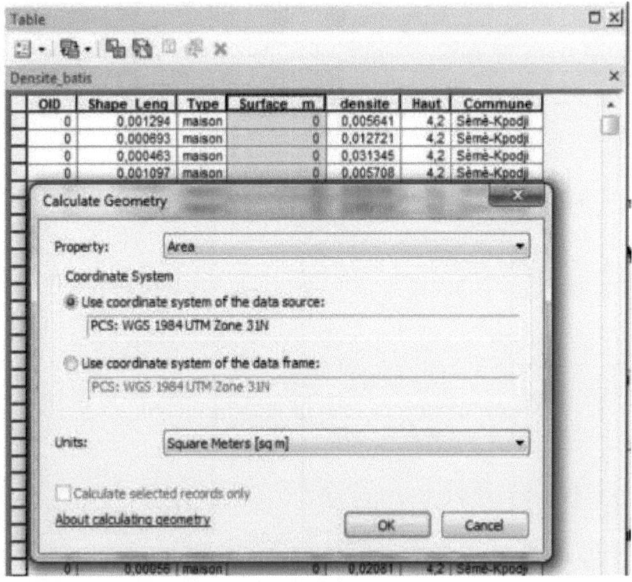

Table 11: Calculation of the surface area of buildings

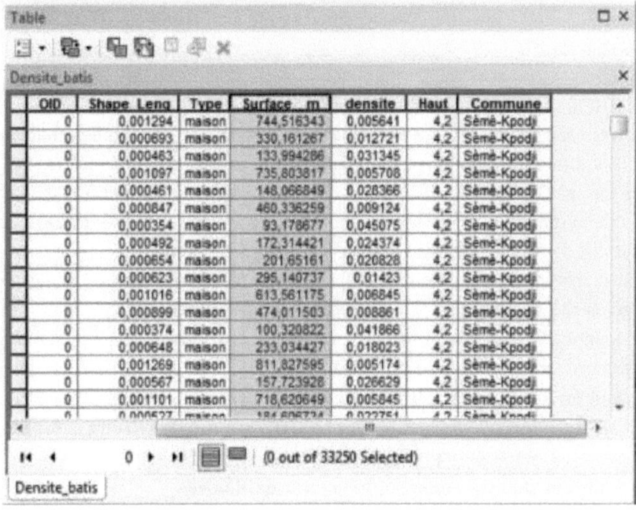

3.4.2.1.5. Density mapping

The attribute tables that provide information on the built-up areas of each commune are сотрlĕ!ёез by two other "fields" that indicate the coordinates of the centroids. Each polygon is then transformed into a point integer whose coordinates are those of the centroids. A spatial join is performed in order to combine each point with the area of the polygon from which it derives.

Referring to the dëlimitation parcellaire dëfmit par le cadastre, there are approximately 70 lots/кт^2 . In fact, a lot dëlimits an area of 200m x 50m with a 15m wide right of way all around which gives 215m x 65m = 13,975m2. Knowing this area, we can calculate the number of lots

59

per Km2. Figure 12 illustrates this dëlimitation.

NB : In order to assess the building density in this study, we were interested in the number of lots per km2 (Iot/Km2). The height of the buildings was not taken into account due to the

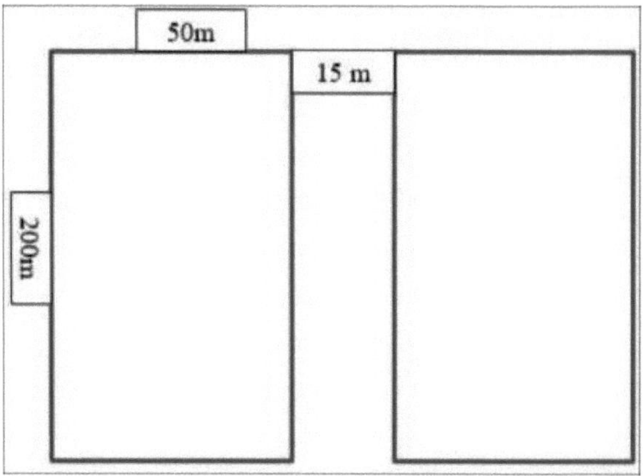

diificulties encountered.

An analysis of figure 12 shows the area occupied by the total number of lots that can be observed on $1km^2$, i.e. :

Number of batches $= \dfrac{1000\,X1000}{13975}$.

This report was used to identify the lot/кт^2 density thresholds for the areas concerned (urban and rural). This formula is implëmentëe in the Excel spreadsheet alongside the attribute fields that indicate the spatial dimension. At this stage, we have a three-column file that allows us to move on to three-dimensional mapping. The result in a GIS software package leads to an assessment of the density distribution of the six communes ëtudiëes.

Furthermore, it should be pointed out that this mëthode has not etc applied in the commune of Aguegues given the вреайеке of the habitats which are pile structures. Here there was etc question of counting the number of habitats per surface area. This is for the simple reason that the land register in terms of subdivisions is not renuirque.

3.4.3. Method for analysing the influence of green spaces on UHIs

To demonstrate the influence of green spaces on UHI, approaches relating to the estimation of carbon storage in aerial biomass are used.

Figure 22 : Parcel boundaries defined by the land register

3.4.3.1. Field measurement of living aerial biomass

In order to measure the living aerial biomass in the field, we delimited 11 square plots of 20 m x 20 m in the conservatory (2.2 ha) of the botanical garden, considering a sampling rate of 20% (Dagnelie, 1998).

In fact, by applying this 20% sampling rate to the total surface area of the conservatory, we were able to obtain the surface area to be sampled (SE) Equation 2).

$$\boxed{\text{SE} = \text{St x Ts}} \qquad (2)$$

SE : Area to be sampled (0.44 ha)
St : Total surface area of the conservatory (2.2 ha)

Ts: Polling rate (20%)

The number of square plots to be considered has etc dëterminë from liquidation 3. Indeed, if n is the number of square plots and a square plot has an area of Sm^2 , then we have :

$$n = \frac{SE}{Sm^2} \qquad (3)$$

n: number of plots

SE : Area to be sampled

Sm^2 : Area of a square plot

All trees were measured in each plot. The following variables were noted: spacing, circumference, total height and tree condition. The plots are located in the fagon ateatoire conservatory. Figure 13 shows the distribution of trees and plots in the JPN conservatory.

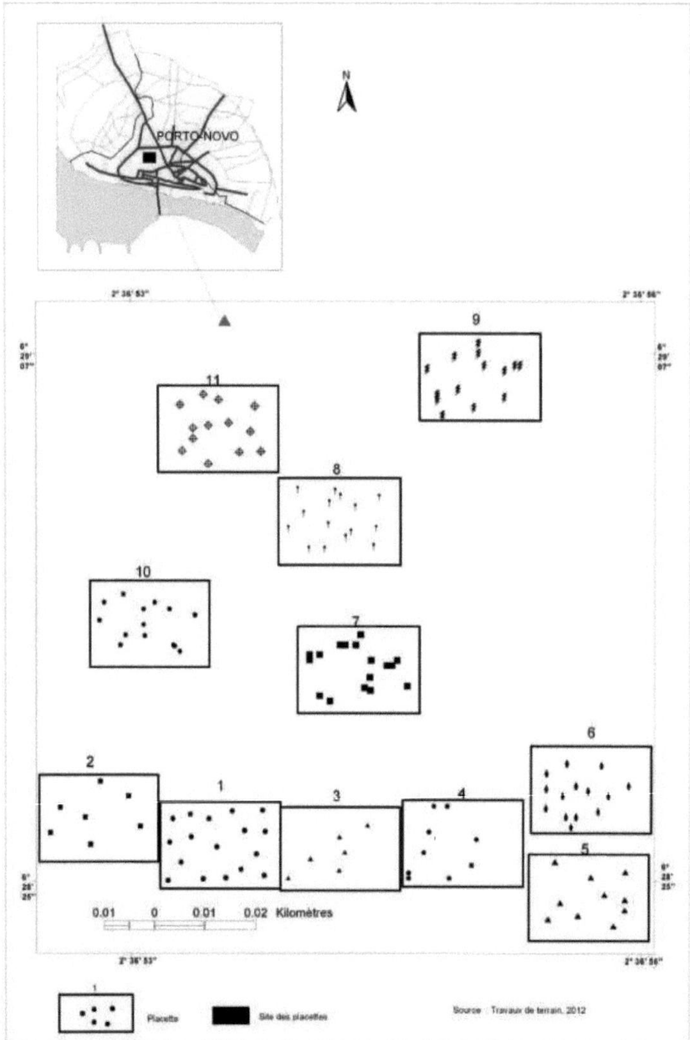

Figure 23: Rëpartitions of trees and plots in the JPN conservatory.

> Estimation of biomass

Many authors nowadays recommend the dëvelopment of specific ёдиайопз regression for the espëces most encountered in etuclics atm sites to amëimprove the йаЫШё of the results (MacDiken, 1997; IPCC, 2003). To estimate the biomass of large trees, it has etc recommandë to use liquation from FAO (1997).

$$Y = \exp(-1,996 + 2,32 \ln D) \qquad (4)$$

Y= biomass per tree in kg; D = diamëtre (dbh) in cm.

Thus, the sum of the biomass of all the trees in a plot gives the total biomass of that plot. So we have :

62

$$Yp = \sum Y \qquad (5)$$

Yp: total biomass of a plot

Y: tree biomass

Then, using liquidation 6, we estimated the average biomass of a plot by calculating the arithmetic mean.

$$Ym = \frac{\sum Yp}{n} \qquad (6)$$

Ym: average plot biomass

Yp: total biomass of a plot

n: number of plots

Similarly, from the average biomass of a plot, we were able to deëduce the total biomass of the conservatory using a rule of three Equation 7).

$$Yt = \frac{S * Ym}{SE} \qquad (7)$$

Yt: total biomass of the conservatory

S: total surface area of the conservatory (2.2 ha)

Ym: average plot biomass

SE: Surface area to be sampled (0.44 ha)

Similarly, the total biomass of site 2 (detente), has *been* reduced by liquation 8 :

$$Y2 = \frac{S2x\ Yt}{s} \qquad (8)$$

S2: area of site 2 [relaxation (1.6 ha)].

Y2: total biomass of site 2

Yt: total biomass of the conservatory

S: total surface area of the conservatory (2.2 ha)

J **Estimating the carbon stock in aerial biomass**

To estimate the carbon stock in the Adriatic biomass, the following measures *were* studied:

- dendrometric measurements (tree circumference and height) taken in the field
- estimation of aerial biomass.

These measurements were only taken in the botanical gardens of Porto-Novo, more specifically in the conservatory or plant museum (site 1), and only took into account all the trees in each of the plots.

- **Estimation of carbon stock in aerial biomass**

The carbon stock in Adriatic biomass was estimated using equation 9 (IPCC 1996). Due to time and financial constraints, the fraction of carbon in biomass used will be a default value proposed by the Intergovernmental Panel on Climate Change (IPCC). The default carbon fraction is 0.5.

$$CE = B \ x \ FC \qquad (9)$$

EC = carbon stored (t C/ha); B = biomass (t/ha); FC = carbon fraction %).

3.4.4 Method for characterising UHIs, their impact on the environment and the population, and proposed solutions for sustainable regulation of UHIs (OS3)

The aim here is to identify and then characterise urban heat islands. At this level, the aim is to show that thermal distribution depends essentially on topography, building density and land use. In other words, the aim is to link the distribution of temperatures to the characteristics of the built-up area (Qudnol et *al*, 2007). The following method was used to demonstrate the relationship between building density, land use and temperature distribution within the city of Porto-Novo and the surrounding area.

In this way, heat islands *have been* characterised in the study area by deducing three main zones: low UHI zone, medium and/or high UHI zone. In each of these zones, urbanisation and population density are characterised. This made it possible to compare these two aspects in each zone and to test hypothesis 2.

3.4.4.1.   Locating and identifying urban heat islands ICU

Several studies use tëlëdëtection to find where ICUs are located. Most of the studies carried out in Montreal use this technology (**Achour, 2006**). Image mapping analysis, based on maps, can be used to determine the surface temperature of a city and to determine the type of soil cover (Aniello *et al.*, 1995; Cavayas and Baudouin, 2008; Gill *et al.*, 2008). Inexpensive and practical, it also allows ë the juxtaposition of maps containing different data from the urban ëШШë environment (soeioeeononiic data, forest cover, etc.) (Quenol *et al.*, 2007). In contrast, other ëtudies use data from mëtëorological stations, tempërature sensors, mobile thermomëtres, to identify UHIs (**Efe S.** *and A.* **Eyefia, 2014, Xavier F., 2015)** and there are cost and logistical issues involved in obtaining measurement sites with daily monitoring.

According to authors such as Williams and Smith (1989); Aniello *et al.*, 1995; Cavayas and Baudouin (2008); and Gill *et al.*, 2008, there are satellite sensors that operate in the thermal infrared to dëtect surface tempëratures. In the present ëtude, we have ийНзë the mëthode of tëlëdëtection to ëstudy UHIs.

3.4.4.2.   Using MODIS images to collect UHI data

The use of MODIS images is based mainly on the analysis of tempërature data, from thermal satellite images, colleges over the city of Porto-Novo and its surroundings, in order to facilitate the study of the phënomëne of urban heat island.

Thus, we opted for tëlëchargeable online satellite images from mëtëorological satellites. These often include sensors that probe l'atmospliere for repërer cloud structure and mëtëorological systëmes, the effects of urban heat islands, ocean currents, phënomënes such as El Nino, forest fires, volcanic ëт188юп8 and those of polluting industries.

After making an inventory of the various sensors that provide appropriate data for our study, we opted for the American MODIS sensor (Moderate Resolution Imaging Spectroradiometer) installed on the Terra (since 1999) and Aqua (since 2002) satellites (Wan Z. et *al.*, 2002). There are several reasons for this preferential choice:

> free MODIS imagery, easily tëlëchargeabIe ;

> The MODIS sensor has a range of 2330 km, enabling it to observe every point on the Earth every one to two days in 36 spectral bands (visual, near-IR, mid-IR, thermal IR) with wavelengths ranging from 0.4 to 14 pm (Chu et al., 2002). Indeed, the 36 co-registered and precisely ëcalibrated spectral bands provide information on land and ocëan surface tempërature, primary productivity, land cover, clouds, aërosols, water vapour, tempërature profiles, and fires.

> The advantage of the high temporal resolution of MODIS is that it is possible to build up image series in order to analyse seasonal or phënological behaviour (Reed et al. 1994);

> good spatio-temporal, radiometric and spectral resolution ;

> high temporal resolution (*one day pass and one night pass made by each satellite every 48 hours = 4 observations in 2 days*). (Petrenko & Ichoku, 2013).

> good spatial resolution, which enables very large portions of the Earth to be covered with good accuracy (varying between 250 m and 1 km) (Chu et al., 2002);

> The best compromise between high temporal resolution and optimum spatial resolution,

which represents a surface precision six (6) times finer than that of the sensor specially designed to study surface temperature, i.e. SPOTVEGETATION, or even NOAA-AVHRR or ENVISAT-AATSR (Advanced Along Track Scanning Radiometer) (a resolution of 3 Km to 5 Km).

> The frequency of passes over the same point for the MODIS satellite is twice daily, whereas it is 35 days for AATSR and daily for AVHRR. Since MODIS is based on two satellites, Aqua and Terra, it is possible to obtain data up to four times a day, and the combination of these two satellites gives an average daily LST that is closer to the ground truth. The local transit times for these two satellites are provided in the product tëlëchargë. In ascending mode, Terra and Aqua pass over North Amërica in the mornings and over North America in the evenings in descending mode.

> offers a network of ground control points bringing the accuracy of geolocation to 45m (geo rdfdrencement), instead of 150m. This gives a probability of correlation between the pixel and the terrain of 65%, compared with 13% without the GCP (Ground Control Points) network. This probability increases with the number of points taken into account. This is why, despite everything, for the purposes of statistical calculations or transects, it is necessary to choose pixel groups of at least 3x3, if not 5x5. In this case, the moving values obtained are significantly improved and much closer to reality, although this method does affect the resolution.

> high-precision gëomëtric and atmosphëric corrections ;

These high-quality bounces (Wan Z. *et al.*, 2002) are available free of charge from NASA's Land Processes Distributed Active Archiv Center (LP DAAC).

In addition, we have opted for the sensor embedded in the TERRA (MODIS-TERRA) satellite platform (EOS - Earth Observation System), as it enables us to monitor l'ëlяl of the Earth's environment and the gradual changes in its climate. Its 'Aqua' counterpart is used to study the water cycle on the Earth's surface and in the atmosphere.

For our ëtude, we chose to use 8-day composites "LST 8-days 1 km" (optimal frame). In fact, the sensor 'photographs' and measures the same land surface almost once a day on average (2 times a day). But the reflectances obtained (particularly in the wavelengths used for surface temperature) are automatically analysed over a series of 8 days, and the least noisy (i.e. obscured by clouds) is selected.

This process corrects the effects of atmospheric reflectance on the image. Geometric corrections are also applied to make the image usable more quickly.

In addition, the LSTs measured by the MODIS sensor are also improved thanks to a new algorithm, the day-night split-window, which takes into account the variation of emissivities as a function of time, measured in seven thermal infrared bands, i.e. bands 20, 22, 23, 29, 31, 32 and 33 (Wan *et al.*, 2002b). MODIS has etc. calibrated on Lake Titicaca (Wan *et al.*, 2002a).

3.4.4.3.   Steps for acquiring and downloading images

There are several ways of accessing MODIS data: REVERB, EARTH EXPLORER, GLOVIS, LP DAAC Data, Pool, MRTWEB...See list and access in https://lpdaac.usgs.gov/get data, consulted on 10/08/2016.

We have selected one of the easy-to-use interfaces for downloading which is REVERB (NASA/EOSDIS): http: // reverb.echo.nasa.gov/reverb, 10/08/2016.

MODIS data are stored in areas equivalent to 10° longitude by 10° latitude. The study area lies between 2°24 and 2°44 East longitude and 6°21 and 6°40 North latitude.

To access the data, set the following parameters

-    **location**: insert the coordinates of the area being searched or simply the
name of the place searched (Place, Name);

-    **the product**: enter the code (see list of MODIS products) in the "Search Terms" section:
example MODIS LST 8-Days 1 km or MOD13Q1 for Terra vegetation indexes, 250m, 16
days;

-    **1 time interval**: start and end dates of the required period (January 2001-December
2001, ... January 2015-December 2015).

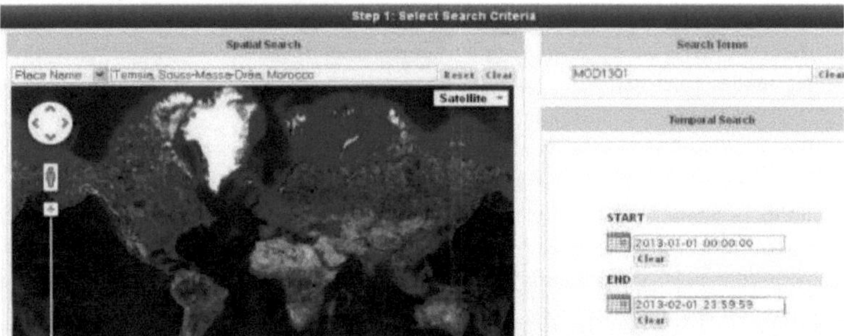

Figure 24: Criteria selection stage

Lktape 2 allows you to зёксйоппег the types of products available:

Figure 25: зёксйоп step for data or product types

Step 3 produces the "granules", i.e. the available data, in the form of a list or map.
From the list of "granules", select the data to be downloaded, and put them in the basket.

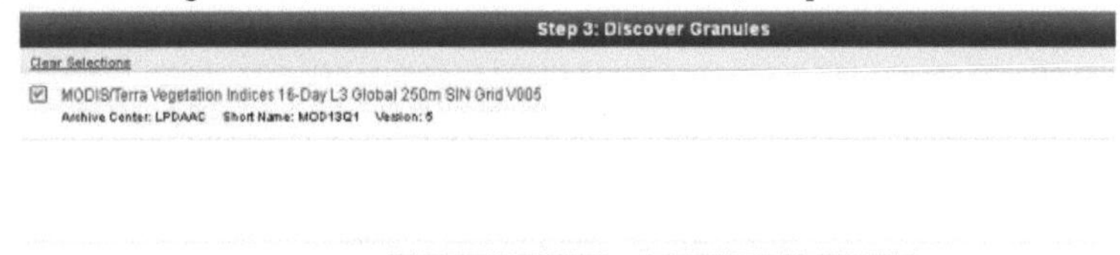

Figure 26: Step of зёксГкп scenes Second tёlёloading option.
Së select the "granules" to be tёlёcharged

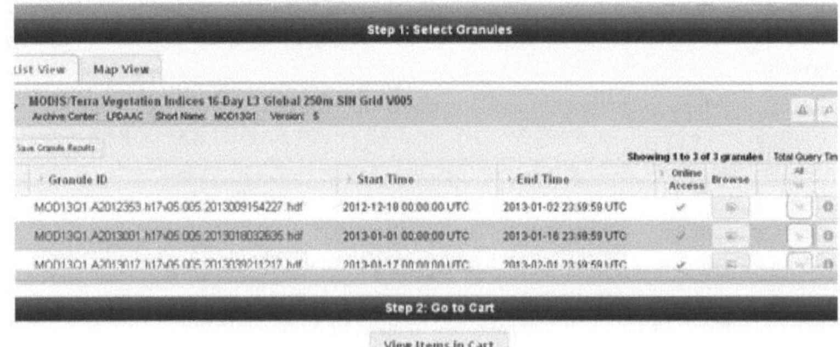

Figure 27: Granule selection stage

See selected granules :

Figure 28: Granule loading stage

Finally tdldownload the files (choose from the 2 proposed tdldownload options. Moreover, to conclude this research, it was necessary to enter the desired period for a annëe, then rëpëter foperation fifteen (15) times (2000 a 2015). Indeed, in this ëtude the period chosen in cows of each year is the great dry season between the months "December and March". This period was chosen because it will allow cloud-free images to be obtained in order to better identify UHIs.

Once the search was launched, all the files were displayed and downloaded. We then proceeded to retrieve the detection data via the LPDAAC site, using MODIS Terra (EOS) data. This method of accessing the data series is fairly straightforward and easy to use. Finally, HDF-EOS format files are downloaded using the WGET command from a Linux Unbutu terminal.

We have therefore gathered a very large number of composite 8-day images corresponding to the dry period (December - March) of the 15 years from 2001 to 2015 described above.

3.4.4.4. Administration of questionnaires to the population and measurement of temperatures at neighbourhood level

> **Administration of questionnaires to the population**

At this level, questionnaires were administered to the population using the sampling method previously defined. Similarly, an interview guide was also administered to local players (town hall staff, neighbourhood chief, etc.).

The questionnaires and interview guide were used to gather information on people's perceptions of the heat, how they felt about it and the impact of the excess heat. This method also enabled us to understand the steps taken and policies adopted by local players, in particular the town hall, to mitigate the excess heat lost by the population.

> **Taking temperature measurements at district level**

Measurements of 1етрёгаШге8 a I echelle de quartier ont ё!ё prises a 1 aide des appareils de mesures de 1етрёгаШге ambiante et de surface infrarouge. The sëries of measurements were ё!ё prises 81тикапётёп1 a la même heure (entre midi et 15H30) au niveau des diflerents types d'espaces bien зрё^щ^з : town centres, crossroads, traffic lights, shopping areas, roadsides, areas with trees or vëgëtaux, near the lagoon, in traffic and areas rich in vëgëtation such as urban green spaces.

These tempërature measurements made it possible to put a figure on how people felt during the përiode of heat dëfmie (midday and 3pm and into the night) by this lastiëre. This is how the maximum and minimum tempërature that could be obtained were measured. The measurement sites were chosen randomly from among the areas listed above.

In addition, four (4) devices (mobile probe thermometer) were ё!ё тоbІH3ё3 for these measurements. However, in order to obtain a certain number of variable points, we carried out^ two sëries of measurements in the same day and this, over five (5) days including a total of eight (8) sites. The table below illustrates these series of measurements.

Table 12: Sëries of tempërature measurements

1ᵉʳᵉ series of measurements					
Sites Days^^	Site 1 JPN	Site 2 Dense crossroads with traffic lights tricolour	Site 3 Lagoon bank	Site 4 Central running (Ouando)	Timetable 12H-13H30'
Day 1	T°max T°min Average temperature	T°max T°min Average temperature	T°max T°min Average temperature	T°max T°min Average temperature	T°max T°min Average temperature
Day 2	T°max T°min Average temperature	T°max T°min Average temperature	T°max T°min Average temperature	T°max T°min Average temperature	T°max T°min Average temperature
Day 3	T°max T°min Average temperature	T°max T°min Average temperature	T°max T°min Average temperature	T°max T°min Average temperature	T°max T°min Average temperature
Day 4	T°max T°min Average temperature	T°max T°min Average temperature	T°max T°min Average temperature	T°max T°min Average temperature	T°max T°min Average temperature
Day 5	T°max T°min Average temperature	T°max T°min Average temperature	T°max T°min Average temperature	T°max T°min Average temperature	T°max T°min Average temperature
Total	T°max T°min Average temperature	T°max T°min Average temperature	T°max T°min Average temperature	T°max T°min Average temperature	T°max T°min Average temperature
2ᵉᵐᵉ series of measurements					
Sites Days"^^	Site 5 In traffic (Porto-Novo city centre)	Site 6 In traffic (Adjarra area)	Site 7 Green spaces (town centre)	Site 8 Green spaces (periphery)	Timetable 1.30-3.15 PM
Day 1	T°max T°min Average temperature	T°max T°min Average temperature	T°max T°min Average temperature	T°max T°min Average temperature	T°max T°min Average temperature
Day 2	T°max	T°max	T°max	T°max	T°max

	T°min Average temperature	T°min Average temperature	T°min Average temperature	T°min Average temperature	T°min Average temperature
Day 3	T°max T°min T°average	T°max T°min T°average	T°max T°min T°average	T°max T°min T°average	T°max T°min Average temperature
Day 4	T°max T°min Average temperature	T°max T°min Average temperature	T°max T°min Average temperature	T°max T°min Average temperature	T°max T°min Average temperature
Day 5	T°max T°min Average temperature	T°max T°min Average temperature	T°max T°min Average temperature	T°max T°min Average temperature	T°max T°min Average temperature
Total	T°max T°min Average temperature	T°max T°min Average temperature	T°max T°min Average temperature	T°max T°min Average temperature	T°max T°min Average temperature

Source: Fieldwork 2017

This table shows that the first series of measurements was taken from 12:00 to 13:30 and the second series from 13:45 to 15:15, with updates every 15 minutes in order to see the different variations during these time periods. For each series of measurements, temperatures were taken simultaneously at the four (4) sites.

3.4.4.5. Trials to propose solutions for sustainable regulation of ICU

The approaches used to find solutions for controlling and regulating CUI are based on cross-analyses (Figure 27) of the data and results from the studies carried out in relation to the first three objectives.

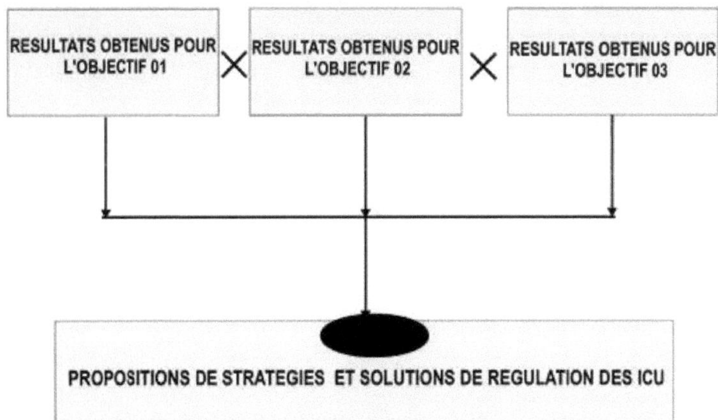

Figure 29: Schëma of analyses cro13ëe8 for the UHI att^nuation proposal.

The choice of 1 analysis cro!3ëe of the results stemming from objectives 01, 02 and 03 made it possible to dëfmir strategies for the att^nuation of ICU in the study environment. Indeed, the ëlëments relating to the options of amënagement and the impact of ICU on l'environnement have ë!ë ийНзёз for l'atteinte de cet objectif.

Partial conclusion

Several mëthodological approaches are used in this ëtude, and in most cases make rëfëference to authors who have already ийНзёз. Elies are chosen according to each objective, and have made it possible to collect and analyse data relating to the paramëtres that explain UHI variation, to analyse the influence of built deeisite and green spaces on UHI variation. Elies have also made it possible to determine the impact of these UHIs on the population and the environment through their identification and quantification, in order to propose an appropriate solution for the sustainable regulation of UHIs. Après avoir posë les bases nietliodologiques, la deuxiëme partie de cette ëtude présente les résultats et discussions dëclinës en quatre chapitres suivant les objectifs.

PART 2

RESULTS AND DISCUSSION

CHAPTER IV: LAND USE DYNAMICS AND POPULATION DENSITY

In this chapter, it is prёзеп!ё on the one hand the results relating to Involution de 1 occupation du sol de 1972 a 2012 and on the other hand the proposal of a зсёпапо future for the sustainable management of 1 space. In fact, this part made it possible to dёgager the trends in urbanisation of 1 space and the dynamics of mineralisation of surfaces as a function of dёmography.

4.1. Changes in land use

The dynamics of land use involution are lost through the spatialisation of the land use units (LUs) presented by the 1972, 1992, 2012 and 2032 maps (from the simulation) and through their respective statistics.

Lntude de revolution des ипкёз d'occupation du sol est fondёe sur trois cas de figure. These are "modifications", "conversions" of these ипкёз which are opposed to "no change" situations. By "modification" we must understand the changes that have taken place within the same catёgory of land use (whether ga is a Rёgression or a Progression). Whereas 'conversion' is the change from one catёgory to another. The term 'unchanged' refers to the set of classes that have remained in the same class between different study dates, i.e. having been affected neither by modifications nor by conversions.

4.1.1. Land use in 1972

At this date, the study area is dominated by natural and man-made formations. Figure 28 shows the Land Use Units (LU).

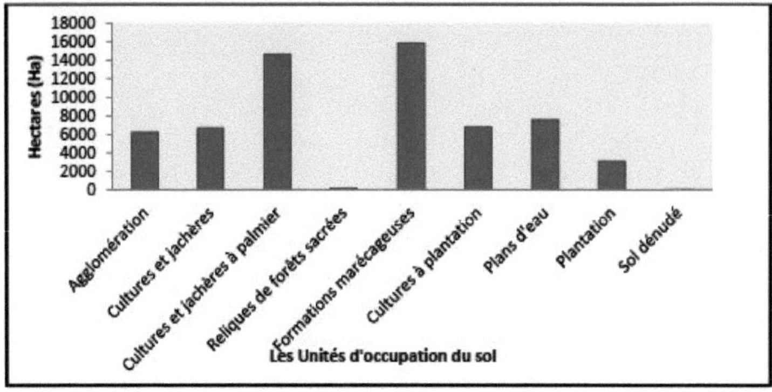

Figure 30 Evolution of UOS in 1972

Source: Data processing, 2017

Figure 29 shows that there are nine (09) UOS. It is ретащиё that in 1972, marecageous formations cover the largest part of the region with an area of 15812.78 hectares, or 25.94% of the study area; Next come palm crops and fallow land covering 14,634.14 hectares (24.01%), followed by bodies of water (12.45%), plantation crops (11.10%), crops and fallow land (10.99%) and, lastly, agglomerations covering 6,235.05 hectares or 10.23% of the total area. Figure 31 shows the status of UOS in 1972.

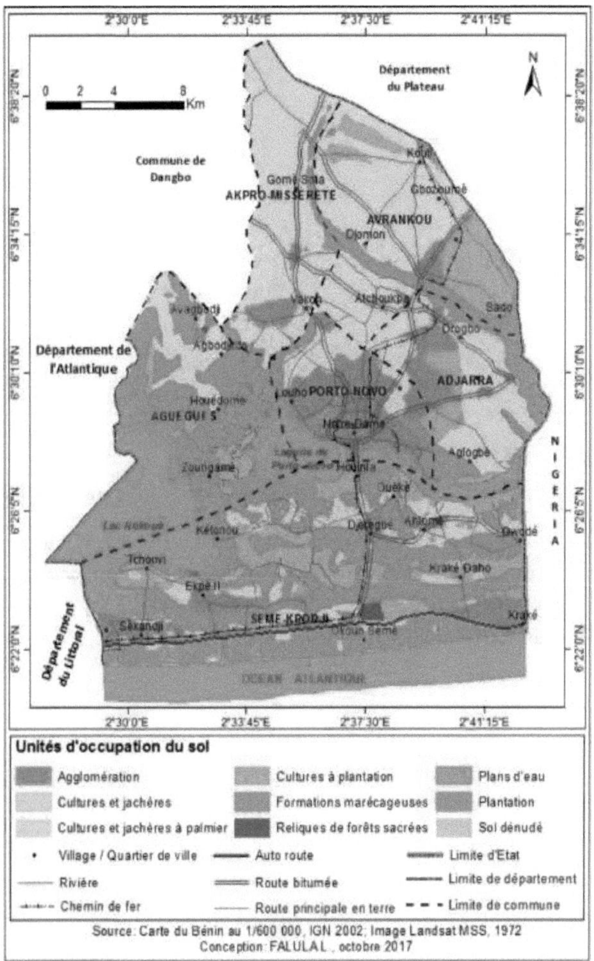

Figure 31: Land use in 1972

Observation of Figure 29 shows that the areas of spontaneous vegetation are in particular: the Sacred Forest Relics and the plantations which together make a percentage of 5.26% on the total area of the study environment. Similarly, with regard to this figure, it is сonз!alë I^gale repartition des taches correspondantes au bati (agglomëration) qui fait un pourcentage de 10,23 %. Considering the latter percentage the non-bati thus occupies 89.77%.

4.1.2. Land use in 1992

With regard to the state of land use in 1992, figure 30 shows relatively the same distribution as in 1972, i.e. a predominance of the following classes: crops and palm groves, crops and picliere, plantation crops, marecageous formations, water plants and agglomerations.

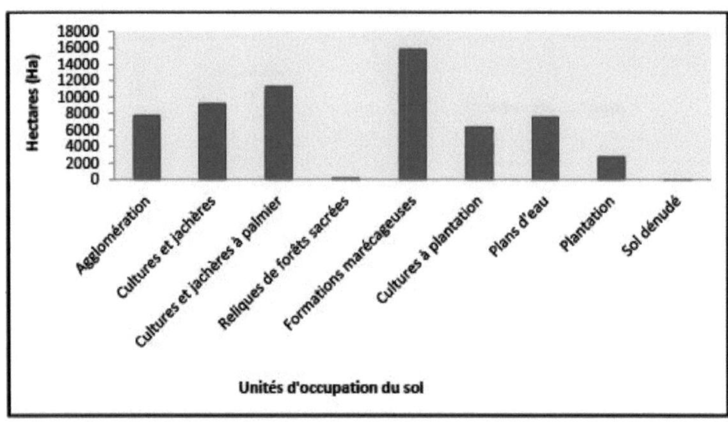

Figure 32: Changes in UOS in 1992
Source: Data processing, 2017

An analysis of figure 31 shows that wetland formations are still in the lead with a surface area of 15833.74 hectares, i.e. 25.97% (slight increase). Next come palm crops and fallow land with 11292.98 hectares or 18.53% (down), water bodies with 7578.92 hectares or around 12.43% and plantation crops with 6361.63 hectares or 10.44%. Crops and fallow land cover an area of 9173.11 hectares, i.e. 15.05% of the total area, followed by agglomerations covering 7755.25 hectares, i.e. 12.72%.
Figure 32 shows the status of UOS in 1992.

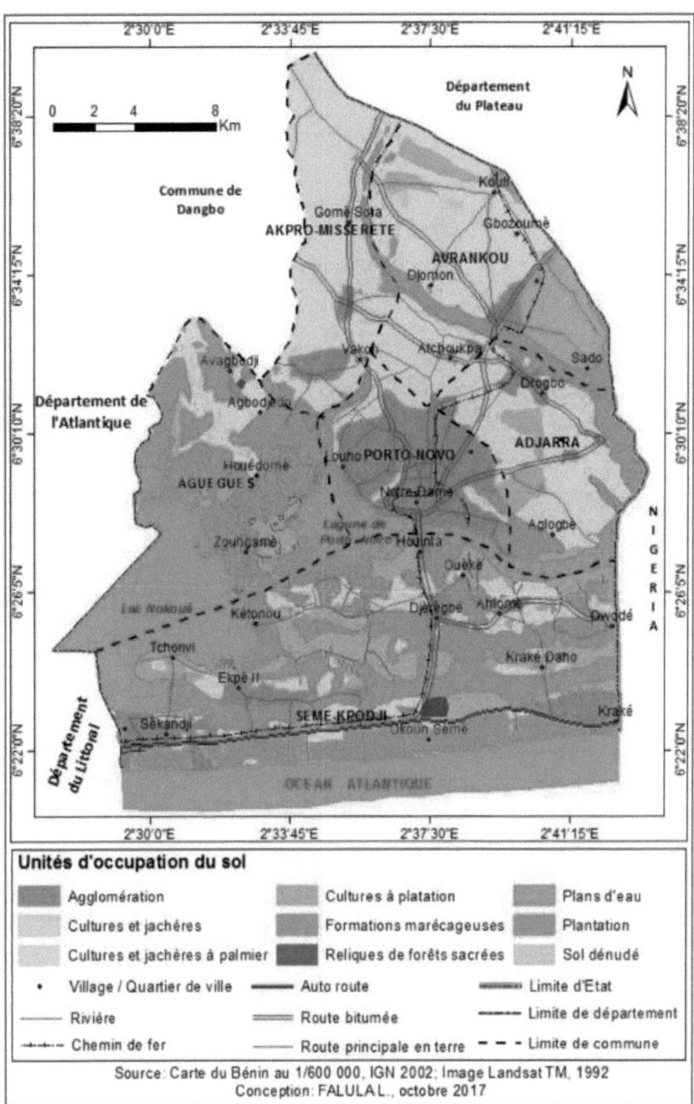

Figure 33: Land use in 1992

As in 1972 the trends in 1992 show ёдалетеп1 that the areas of spontaneous уёдёlайоnз still remain the relics of sacred forests and plantations which together make a percentage of 4.74% (decreasing) on the total area of the study environment. Similarly, the even distribution of built-up areas (agglomëration) is still observed, with a surface area of 7755.25 ha compared with 6235.05 ha in 1972, an increase of 124.38 %. From this point of view, undeveloped land occupies 87.28% of the total, a decline of 2.49% over almost two decades.

4.1.3. Land use in 2012

In 2012, the UOS observed previously are still present but, this time, with a significant increase in the surface areas occupied by agglomerations and a regression in the number of UOS.

Considĕrable of natural formations. Figure 33 shows this ёyo1ийоп.

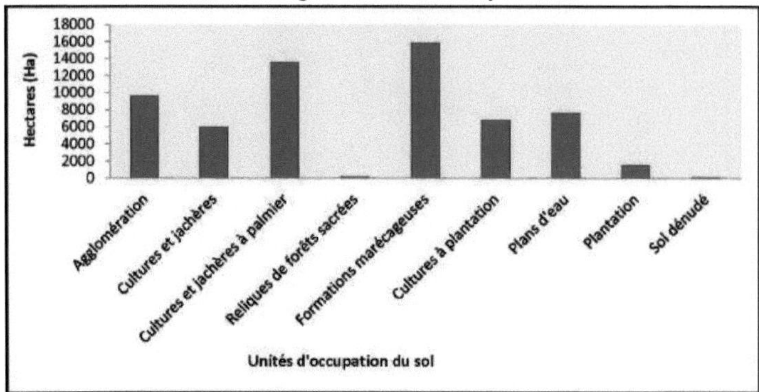

Figure 34: Change in UOS in 2012
Source: Bounces processing, 2017

A reading of the observed frequencies shows that the moorland formations continue to occupy a very large area, i.e. 15824.48 hectares (25.96%). Crops and palm groves cover an area of 13,527.04 hectares (22.19%), while agglomerations cover an area of 9,581.01 hectares (15.72%).

Figure 34 shows the status of UOS in 2012.

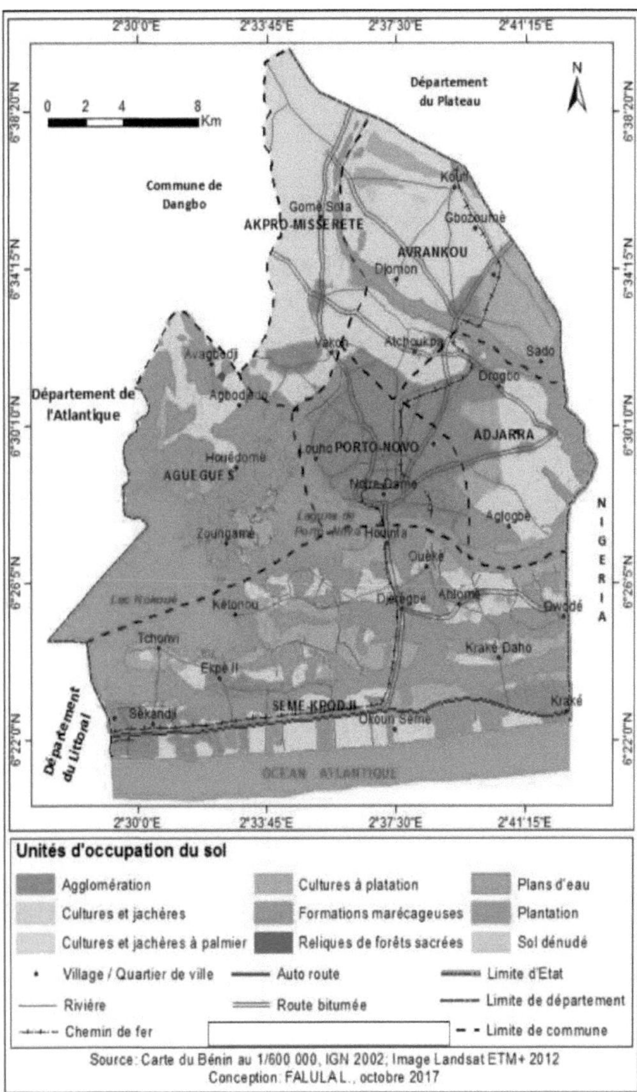

Figure 35: Land use in 2012

Observation of the spatialisation of UOS shows that the built reprёзепlё by agglomёration spots (red colour) occupies more spaces at the dёtriment of natural formations. In this anпёe, the адд1отёгайопз account for 15.72% compared with 12.72% in 1992 and 10.23% in 1972. A remarkable increase has therefore been observed, pushing undeveloped areas backwards as they now account for 84.28% compared with 87.28% in 1972.

4.1.4. Dynamics of land use between 1972 and 1992

Analysis of the dynamics of land use between 1972 and 1992 гёуёк shows that the study area remains partly stable.

However, significant changes that occurred during this përiod mostly concern palm crops and picliere, crops and fallowëre, plantation crops and agglomërations (Table XIII and Figure 35). Such ëyolийоп is Hëc to a strong growth in the region's population.

Table 13: Evoluuion of ипкёз land use between 1972 and 1992

Land use units	Area (Ha) 1972	Surface area (Ha) 1992
Agglomeration	6235,05	7755,25
Crops and fallow land	6700,40	9173,11
Palm crops and fallow land	14634,14	11292,98
Sacred forest relics	140,82	135,08
Marecageous formations	15812,78	15833,74
Plantation crops	6766,45	6361,63
Bodies of water	7590,31	7578,92
Plantation	3068,17	2752,69
Denude soil	11,87	76,59
Total	**60960**	**60960**

Source: Field data

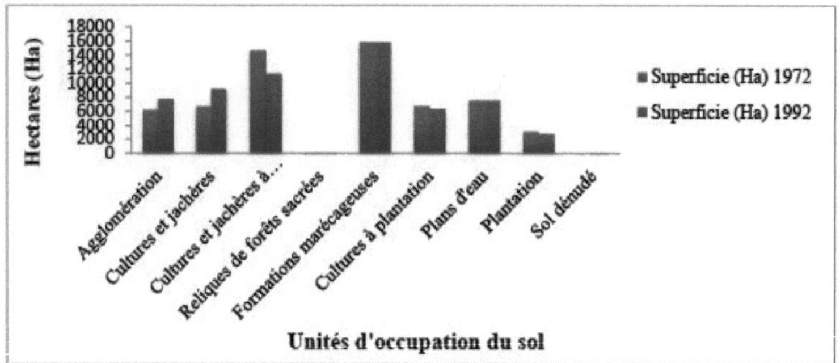

Figure 36: Changes in land use between 1972 and 1992

Source: Data processing, 2017

Looking at this graph, we can say that the most common classes in 1992 are identical to those in 1972. In fact, we can see that :

• crops and fallow land for palm trees currently cover an area of 11292.98 ha or 18.53% compared with an area of 14634.14 ha or 24.01% in 1972. *This represents a decrease of 3341.16 ha or 5.48%;*

• In 1992, plantation crops covered an area of 6361.63 ha, i.e. 10.44%, compared with 6766.45 ha in 1972, i.e. 11.10%. *Hence the decrease in surface area*. It can therefore also be said that this class has changed over the 20 years from 1972 to 1992.

• plantations, which occupied 3068.17 ha (5.03%), fell to 2752.69 ha in 1992 (4.52%). *Hence the decrease in surface area*.

• crops and picliere cover an area of 9173.11 hectares, i.e. 15.05% of the total area, compared with 10.99% in 1972, i.e. 6700.40 ha. *Hence the increase in surface area*. It can therefore be said that this land use unit has changed over the 20 years (1972-1992).

• The agglomerations cover an area of 7755.25 hectares or 12.72% of the study area in 1992. In 1972, they occupied an area of 6235.05 ha or 10.23%. *Hence the increase in surface area*. So we can say that this land use unit has undergone a change over the 20 years (1972-1992). This is shown in Figure 36.

78

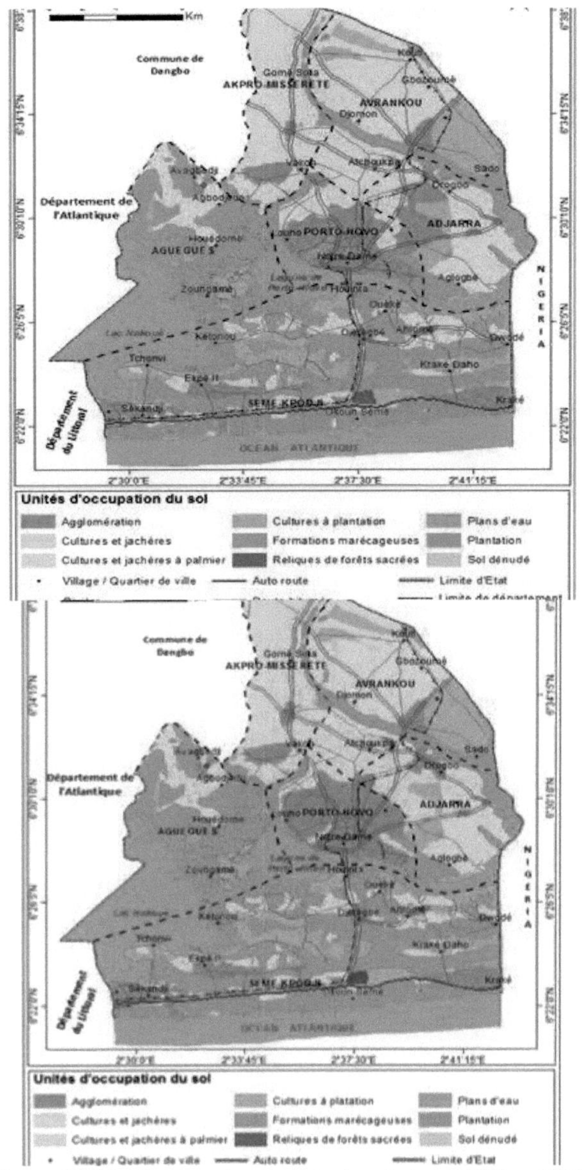

Figure 37: Map showing changes in land use between 1972 and 1992

4.1.5. Dynamics of land use between 1992 and 2012

Between 1992 and 2012, the дёпёгак trend shows an alteration in changes in 1 land use especially in agricultural areas and areas of natural vëgëtation (Figure 36).

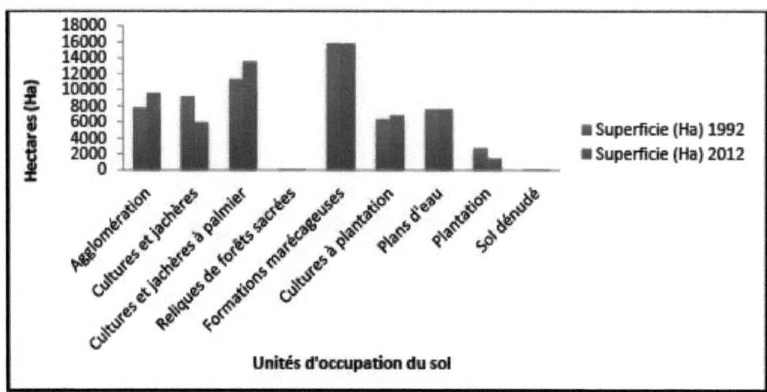

Figure 38: Change in land use between 1992 and 2012

Figure 37 shows that between 1992 and 2012 :

- the agglomëration has *grown* from 7755.25 hectares to 9581.01 hectares
hectares or 3% ;
- cultivation and palm groves increased from 11292.98 hectares to 13527.04 hectares
an increase of 2,234.06 hectares or 3.66%;
- Crops and fallow land *fell back*, losing 3204.17 hectares, or 5.26% of its original area;
- the area of sacred forests has been reduced from 135.08 hectares, or 0.22%, to 95.83 hectares
i.e. 0.16%; *hence the decline*;
- Planted crops *increased by 0.71%* from 6361.63 hectares (10.44%) to 6795.97 hectares (11.15%);
- the plantation fell from 2752.69 hectares to 1483.77 hectares, *a decline of* 2.09%.

As for the other ипкёз land cover namely water body, bare soil and marecageous formations have not known any significant change during the përiode. Figure 38 shows the Involution map of land cover between 1992 and 2012.

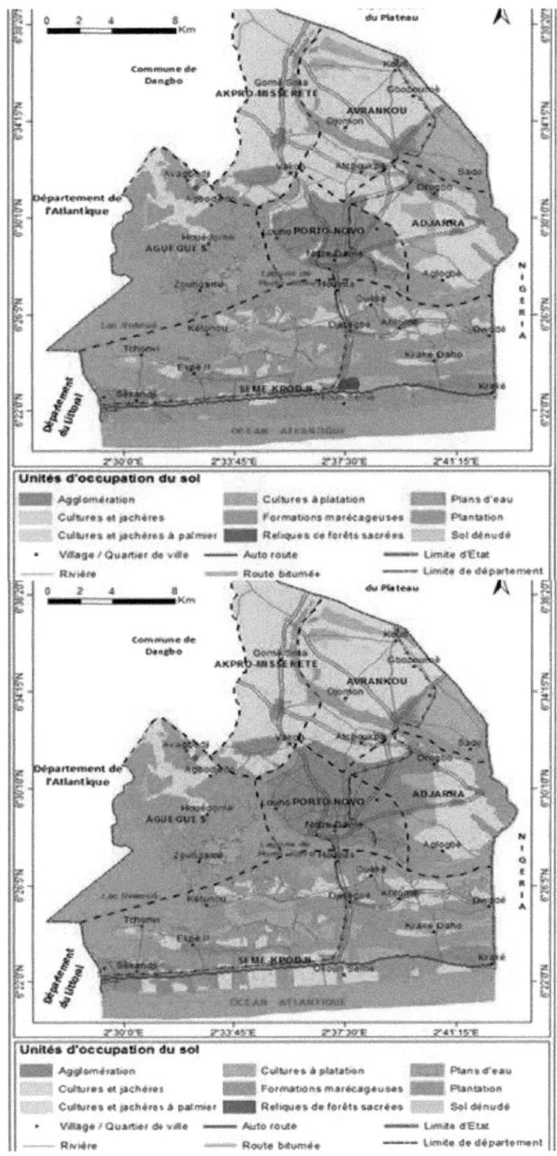

Figure 39: Map showing changes in land use between 1992 and 2012

In view of this evolution, which marks a regressive trend of natural formations to the detriment of anthropogenic formations, it should be noted that agglomërations hold a significant share of transformed spaces. Although the undeveloped areas seem to occupy large areas, the ипкёз that surround them are still affected by the changes observed.

These observations have been taken into account in the simulation exercise that led to the prospective mapping for 2032.

4.1.6. Land use to 2032

In order to map this later land use, the transition matrices are prësentëes with regard to the calculated rates of change and even the mutation probabilities of each UOS.

4.1.6.1. Transition matrix between 1972 and 1992

The Transition Matrix illustrates, in hectare terms, the change in land cover class areas between 1972 and 1992. Indeed, this matrix of changes which is gënërëe by crossing the 1972 and 1992 land use maps shows an evolution at the level of the different ипкёз land use ипкёз (Table XIV).

Table 14: Transition matrix between 1972 and 1992

1972-1992										
	RFS	PL	CJP	CJ	FM	PE	SD	CP	AG	TOTAL 1972
RFS	**135**	0	6	0	0	0	0	0	0	141
PL	0	**2753**	291	0	0	0	0	0	24	3068
CJP	0	0	**10996**	2812	0	0	0	0	826	14634
CJ	0	0	0	**6361**	0	0	0	0	339	6700
FM	0	0	0	0	15813	0	0	0	0	15813
PE	0	0	0	0	0	**7579**	0	11	0	7590
SD	0	0	0	0	0	0	**12**	0	0	12
CP	0	0	0	0	0	0	0	**6351**	415	6766
AG	0	0	0	0	21	0	65	0	**6149**	6235
TOTAL 1992	135	2753	11293	9173	15834	7579	77	6362	7755	**60960**

RFS : Reliques de forets sacrees ; *PL* : Plantation ; *CJP* : Cultures et jaches a palmier ; *CJ* : Cultures a plantation ; *FM* : Formations marecageuses ; *PE* : Plans d'eau ; *SD* : Sol denude ; *CP*: Cultures a plantation ; *AG*: Agglomeration

In fact from reading this table, we can see that of the 141 hectares occupied by the Sacred Forest Relics in 1972, 135 hectares are re3lë8 intact in 1992 and 6 hectares are now occupied by Palm Crops and Fallows. Similarly, of the 3068 hectares occupied by the plantation in 1972, 2753 hectares are re3lë8 intact in 1992, 291 hectares are now occupied by Palm Crops and fallows and 24 hectares by agglomëration. Thus the plantation has lost a total of 315 hectares over a period of 20 years. Also, in 1972, of the 14634 hectares occupied by the Cultures et jachëres a palmier, 11293 hectares are re3lë8 intact in 1992, 2812 hectares are now occupied by the cultures et picliere and 826 hectares by the agglomërations. This represents a loss of 3638 hectares over the 20-year period. Similarly, in 1972, of the 6766 hectares occupied by plantation crops, 6362 hectares are re3lë8 intact in 1992 and 415 hectares are now occupied by agglomërations.

As for crops and yellows, dënudë soil and agglomëration, they each increased by : 2473 hectares or 4.06%, 65 hectares or 0.11%, 1520 hectares or 2.49%.

Average annual rate of spatial expansion (T) or rate of change between 1972 and 1992

Table XV shows the changes that occurred between 1972 and 1992.

Table 15: Change in land use between 1972 and 1992.

Land use units	Area 1972 (Hectares) S1	Surface area 1992 (Hectares) S2	T1 (%)
Agglomëration	6235,05	7755,25	1,091
Crops and fallow land	6700,40	9173,11	1,571

Palm crops and fallows	14634,14	11292,98	-1,296
Sacred forest relics	140,82	135,08	-0,208
Marecageous formations	15812,78	15833,74	0,007
Plantation crops	6766,45	6361,63	-0,308
Bodies of water	7590,31	7578,92	-0,008
Plantation	3068,17	2752,69	-0,543
Dënudë soil	11,87	76,59	9,322
Total	**60960**	**60960**	**9,628**

S1 = area at date 1; S2 = area at date 2; T1 = rate of change between the dates 1972 and 1992.

This table shows that the study area as a whole has **an average annual rate of spatial expansion of "9.628%"**.

In fact, the results of calculating the rate of change or average annual spatial expansion rate between 1972 and 1992 (T1) show that ГЛд1отёгайоп, Crops and Fallow Land, Marecageous Formations and Dënudë Soil present a significant increase with a respective T1 of 1.09%; 1.57%; 0.007%; 9.32% (yellow colour), i.e. more than 4078.60 hectares of growth over a 20-month period. On the other hand, we can see a sharp decrease in the surface area of palm tree crops and fallows, sacred forest relics, plantation crops, water bodies and plantations (green colour).

### 4.1.6.2.	Transition matrix between 1992 and 2012

The Transition Matrix illustrates, in hectare terms, the change in land cover class areas between 1992 and 2012. Indeed, this matrix of changes which is gënërëe by crossing the 1992 and 2012 land use maps of the study area shows an evolution at the level of the different ипкёз land use ипкёз (Table XVI).

Table 16: Transition matrix between 1992 and 2012

1992-2012									
	RFS	PL	CJP	CJ	FM	PE	SD	CP	AG	TOTAL 1992
RFS	**96**	19	0	20	0	0	0	0	0	135
PL	0	**1336**	1050	0	0	0	0	0	367	2753
CJP	0	115	**11048**	125	0	0	0	0	5	11293
CJ	0	0	0	**5824**	0	0	0	3349	0	9173
FM	0	14	0	0	**15819**	0	0	0	1	15834
PE	0	0	0	0	0	**7579**	0	0	0	7579
SD	0	0	0	0	0	0	**77**	0	0	77
CP	0	0	1429	0	5	27	0	**3423**	1478	6362
AG	0	0	0	0	0	0	0	24	**7730**	7755
TOTAL 2012	96	1484	13527	5969	15824	7606	77	6796	9581	**60960**

RFS : Reliques de forets sacrees ; *PL* : Plantation ; *CJP* : Cultures et jaches a palmier ; *CJ* : Cultures a plantation ; *FM* : Formations marecageuses ; *PE* : Plans d'eau ; *SD* : Sol denude ; *CP*: Cultures a plantation ; *AG*: Agglomeration

From reading this table, we can see that of the 135 hectares that the sacred forest relics occupied in 1992, 96 hectares are rе3lё8 intact in 2012, 19 hectares are now occupied by plantation and 20 hectares by Crops and fallow. Thus, the sacred forest relics have lost a total of 39 hectares over a period of 20 years. **This represents a decline of 0.16%.** Similarly, we can see that of the 2753 hectares occupied by the plantation in 1992, 1484 hectares are rе3lё8 intact in 2012, 1050 hectares are now occupied by palm crops and fallows, and 367 hectares by agglom6ration. The plantation has thus lost a total of 1,269 hectares over a period of 20 years, to the detriment of palm crops and fallow land and the agglomeration. **This represents a decline of 2.09%.**

We also note that of the 9173 hectares occupied by crops and set-aside in 1992, 5969 hectares remained untouched in 2012 and 3349 hectares are now occupied by plantation crops. Crops and fallow land have therefore lost a total of 3,349 hectares over a period of 20 years to plantation crops. **This represents a decline of 5.26%.**

Finally, we can see that wetland formations have lost 1 hectare to the detriment of urban areas, which now occupy 15824 hectares in 2012 compared with 15834 hectares in 1992. **This represents a decrease of 0.01%.**

Palm crops and fallow land, plantation crops, water bodies and agglomerations each **increased** over a 20-year period (1992-2012) by : **3,66% ; 0,71% ; 0,05% ; 3%.**

Average annual rate of spatial expansion (T) or rate of change between 1972 and 1992 Table XVII shows the changes in occupancy between 1992 and 2012.

Table 17: Change in land use between 1992 and 2012

Land use units	Surface area 1992 (Hectares) S1	Area 2012 (Hectares) S2	T1 (%)
Agglom6ration	7755,25	9581,01	1,057
Crops and fallow land	9173,11	5968,94	-2,149
Palm crops and fallow land	11292,98	13527,04	0,903
Sacred forest relics	135,08	95,83	-1,716
Marecageous formations	15833,74	15824,48	-0,003
Plantation crops	6361,63	6795,97	0,330
Bodies of water	7578,92	7606,36	0,018
Plantation	2752,69	1483,77	-3,090
Denude soil	76,59	76,59	0,000
Total	**60960**	**60960**	**-4,65**

S1 = area at date 1; S2 = area at date 2; T1 = rate of change between dates 1992 and 2012.

This table shows that the overall **average annual rate of spatial expansion** in the Stude area **is "-4.65%"**.

In fact, the results of calculating the rate of change or average annual spatial expansion rate between 1992 and 2012 (T2) show that the land use units, namely : Agglom6ration, Cultures et jaches a palmiers, Cultures a plantation et Plan d'eau present une importante progression avec un taux T2 respectif de 1,057% ; 0,903% ; 0,330%, 0,018% soit plus de 4521,3 hectares de croissance durant une përiode de 20 anëe3 (couleur jaune).Conversely, we can see a sharp decrease (green colour) in the areas of: crops and jaclieres, sacred forest relics, wetland formations and plantation.

In addition, we note that the "Sol dënudë" occupancy unit has not changed over this përiode with a T2 rate of 0.00%.

4.1.6.3. Transition matrix between 2012 and 2032

The Transition Matrix illustrates in hectare terms the future change in land cover class areas between 2012 and 2032. Indeed, this matrix of changes which is gënërëe by crossing the 1972 and 2012 land use maps of the study area shows an evolution at the level of the different ипкёз land use ипкёз (Table XVIII).

Table 18: Evolution at the level of the various ипкёз land use ипкёз between 2012 and 2032.

	RFS	PL	CJP	CJ	FM	PE	SD	CP	AG	TOTAL 2012
20112-2032										
RFS	16	0	0	80	0	0	0	0	0	96
PL	0	1484	0	0	0	0	0	0	0	1484
CJP	0	35	5988	6476	0	0	0	0	1028	13527
CJ	0	238	0	3437	0	0	0	0	2294	5969
FM	0	0	0	0	15824	0	0	0	0	15824
PE	0	0	0	0	0	7602	0	4	0	7606
SD	0	0	0	0	0	0	77	0	0	77
CP	0	0	0	0	13	0	0	5459	1324	6796
AG	0	0	0	0	0	0	0	0	9581	9581
TOTAL 2032	16	1757	5988	9993	15837	7602	77	5463	14227	**60960**

RFS: Remnants of sacred forests; *PL*: Plantation; *CJP*: Palm crops and fallow land; *CJ*: Plantation crops; *FM*: Wetland formations; *PE*: Water bodies; *SD*: Denuded soil; *CP.* Cultures a plantation ; *AG*: Agglomeration

Looking at this table, we can see that of the 96 hectares occupied by sacred forest relics in 2012, 16 hectares will remain intact in 2032, while 80 hectares will be occupied by crops and heathland, **representing a decline of 0.13%**.

Similarly, of the 13,527 hectares occupied by crops and palm groves in 2012, 5,988 hectares will remain untouched in 2032, of which 6,476 hectares will be occupied by crops and palm groves and 1,028 hectares by agglomeration, **representing a decline of 12.37%**.

In addition, we note that of the 6796 hectares occupied by Plantation Crops in 2012, 5463 hectares will remain untouched in 2032 and 1324 hectares will be оссирёз by аддlотёгайоп8. **Hence a regression of 2.19%**.

On the other hand, we see that crops and piclieres, plantation and адд^тёгайоиз will each experience an **increase** over a 40-year pёriod (2012-2032) of respectively: **6,6%, 0,45%, 7,62%**.

In addition, it should be noted that the relics of sacred forests will almost no longer exist in 2032. In fact, this иmкё of land occupation occupied 95.83 hectares or **0.16%** in 2012 which will become 15.83 hectares in 2032 or **0.03%** of the total surface area of the study area. Hence the trend towards the total disappearance of the sacred forest relics in 2032.

Average annual rate of spatial expansion (T) or rate of change between 2012 and 2032

Table XIX shows the changes in occupancy between 2012 and 2032.

Table 19: Change in land use between 2012 and 2032

Land use units	Area 2012 (Hectares) S1	Area 2032 (Hectares) S2	T1 (%)
Agglomёration	9581,01	14226,97	1,977
Crops and peckers	5968,94	9992,92	2,577
Crops and palm groves	13527,04	5988,46	-4,074
Sacred forest relics	95,83	15,83	-9,002
Marecageous formations	15824,48	15837,21	0,004
Plantation crops	6795,97	5463,22	-1,091

Bodies of water	7606,36	7602,26	-0,003
Plantation	1483,77	1756,54	0,844
Dënudë soil	76,59	76,59	0,000
Total	**60960**	**60960**	**-8,77**

S1 = area at date 1; S2 = area at date 2; T1 = rate of change between dates 2012 and 2032.

This table shows that the study area as a whole is experiencing **an average annual spatial expansion rate of -8.77%.**

Indeed, the results of the calculation of the rate of change or average annual spatial expansion rate between 2012 and 2032 show that the ипкёз of land use namely: Agglomëration, Crops and fallow land, wetland formations and Plantation show a significant increase with a respective expansion rate of 1.977%; 2.577%; 0.004% and 0.844% i.e. an occupation of 41813.64 hectares of the total area compared to 38980.79 hectares in 2012 (yellow colour). On the other hand, there has been a sharp decrease (green) in the surface area of palm crops and fallow land, sacred forest relics, plantation crops and water bodies.

The results also show that the land use unit "Sol dënudë" will undergo no change during the 2012 - 2032 period.

4.2. Simulation of the state of land use in 2032

In order to examine the structure, trend of change and show Involution of 1 land cover in the next twenty ttnnees (20 years), we ёкЬогё the map of 2032 using the modële CA_Markov on IDRISI (Eastman, 2006).

Thus, the prediction of land use in 2032 was made on the basis of the transition between land use in 1992 and 2012. The result of the land use prediction for the year 2032 is illustrated in Figure 39 below.

At this date, land use is dominated by agglomerations, crops and jacliere, palm crops and piclieres, plantation crops, wetland formations and water bodies. Table XX and figure 39 clearly illustrate this state of affairs, as does the distribution of the various themes.

Table 20: Projected evolution of land use units in 2032

2032		
Land use units	**Hectares (Ha)**	**Percentage (%)**
Agglomëration	14226,97	23,34
Crops and workshops	9992,92	16,39
Palm crops and plantations	5988,46	9,82
Sacred forest relics	15,83	0,03
Marecageous formations	15837,21	25,98
Plantation crops	5463,22	8,96
Bodies of water	7602,26	12,47
Plantation	1756,54	2,88
Dënudë soil	76,59	0,13
Total	**60960**	**100**

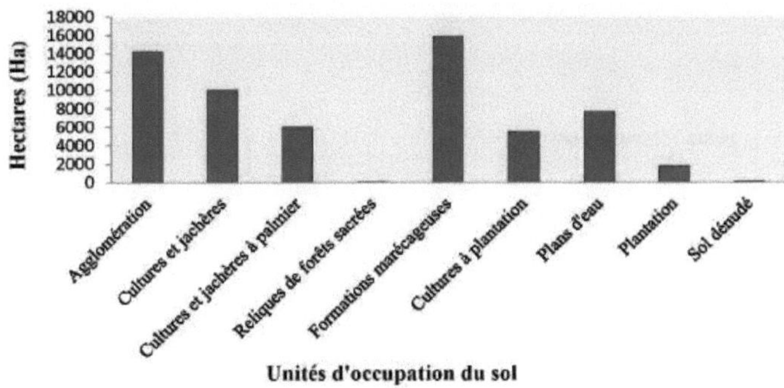

Figure 40: Land use units in 2032
Source: Data processing, 2017

Figure 39 shows that by 2032, land use will be dominated by 6 main themes:
- the built-up areas will cover 14226.97 hectares or 23.34% of the study region.
- crops and fallow land will cover an area of 9992.92 hectares, i.e. 16.39% of the total area.
total surface area.
- crops and fallow land for palm trees will cover an area of 5,988.46 hectares
hectares or 9.82% of the total surface area.
- wetland formations will cover an area of 15837.21 hectares, or 25.98%.
of the study area ;
- crops to be planted will cover an area of 5 463.22 hectares, i.e. 8.96%.
of the total surface area.
- the water bodies will cover 7602.26 hectares or 12.47% of the study area.

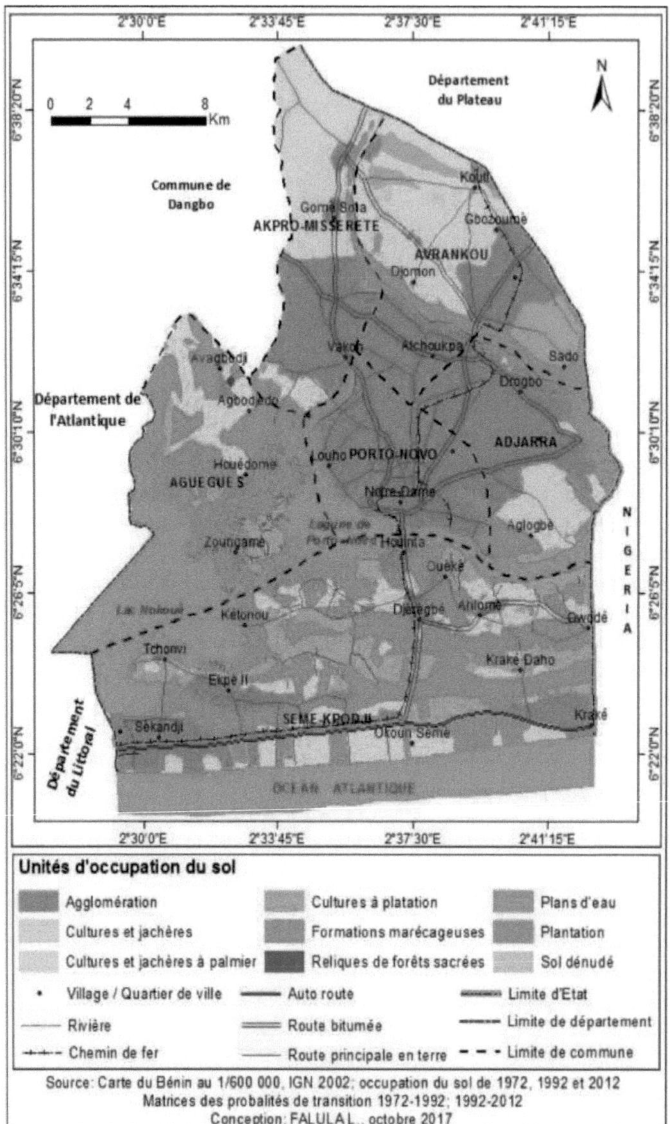

Figure 41: Land use units

Looking at this figure, we can see that the areas of natural vëgëtations are in particular: the relics of sacred forests which together make up a percentage of 0.03% of the total surface area of the study area. Similarly, with regard to this figure, we can say that the built-up area makes a percentage of 23.34% on the total area of the study area. Thus, we can say that the surface area of the agglomération, crops and piclieres, crops and palm fallows, and plantation crops is increasing at the expense of the surface area of the natural vegetation zones.

4.2.1. Simulation of land use dynamics between 2012 and 2032

Between 2012 and 2032, the general trend shows that there will be an acceleration of changes in land use, especially in agricultural areas, natural vegetation areas and in the areas where the land is used for agriculture.

agglomerations (figure 41).

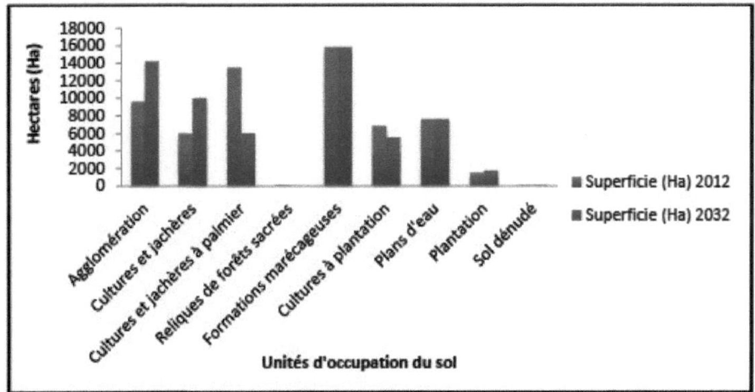

Figure 42: ипкёз land use between 2012 and 2032.

Source: Data processing, 2017

Figure 41 shows that between 2012 and 2032 :

-       The agglomëration will see an **increase of 7.62%** from 9581.01 hectares to 9581.01 hectares.

14226.97 hectares.

-       Crops and pichere will increase from 5968.94 hectares to 9992.92 hectares**, an increase of 6.6%;**

-       Crops and fallow land for palm trees will fall from 13,527.04 hectares to 5,988.46 hectares, **a decline of 12.37%;**

-       sacred forest relics will **decrease by 0.13%** from 95.83 hectares to 15.83 hectares.

-       Crops to be planted will fall from 6795.97 hectares to 5463.22 hectares**, a decline of 2.19%;**

-       The plantation will **grow by 0.45%**, gaining 277.77 hectares;

There will be no change in the wetland formations, water bodies or bare soil over the period.

Figure 42 shows the Involution map of land use between 2012 and 2032.

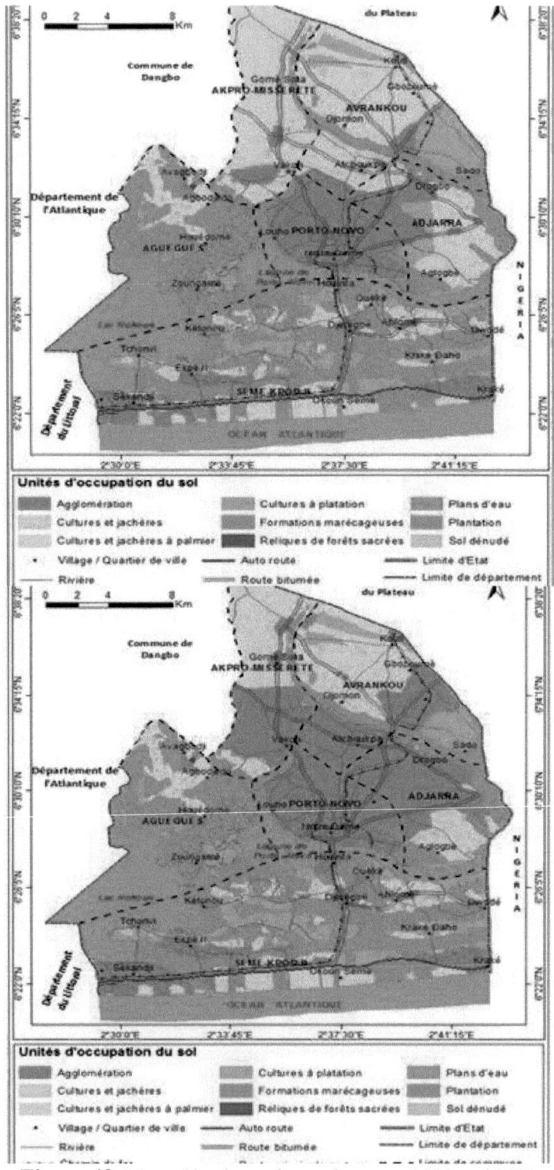

Figure 43: Map of land use change between 2012 and 2032

4.2.2. Summary of land use dynamics

Table XXI summarises the dynamics of land use over the various periods studied.

Table 21: Svntliese of land use dynamics

Përiode UOS	1972-1992	1992-2012	2012-2032
Agglomeration	+	+	+

Crops and fallow land	+	-	+
Palm crops and fallow land	-	+	-
Sacred forest relics	-	-	-
Marecageous formations	stable	Stable	Stable
Plantation crops	-	+	-
Bodies of water	stable	Stable	Stable
Plantation	-	-	+
Denude soil	+	Stable	Stable
Read: - = decrease; + = increase			

This table shows that the vëgëtales formations will disappear if the trend continues. Similarly, cultivated areas have seen an increase with Involution 1 in population numbers. The plant formations and cultivated areas are giving way to agglomerations.

The classes that appear to be the most stable over these various dates are: dënudë soil, water body and wetland formations.

Over the various periods 1972-1992, 1992-2012 and 20122032, areas of natural vegetation (sacred forest relics) disappeared at the expense of agricultural areas (crops and woodpeckers, palm tree crops and woodpeckers, plantation crops) and built-up areas (conurbations). This situation can influence the intensity of UHI.

4.3 Demographic trends and analysis of population density

This section presents the population revolution and the analysis of population density in the research environment.

In recent decades, African cities have experienced considerable demographic growth (Amadou, Klissou et al., 2009). This growth is driving the dynamics of land use (Tribillon, 1992). The commune of Porto-Novo and its surroundings are no exception to this reality.

4.3.1 Population of Porto-Novo and the surrounding area

The results of the census carried out in February 1992 and published in August 1993 show a population of 179,138 for the town of Porto-Novo, which represents approximately 20% of the population of the department of l'Oиётё and 3.6% of the total population of Вёпт. The Atlas monographique des communes du Вёпт publishes a total population in 1999 of 207,190 inhabitants. The Recensement Gënëral de la Population et de l'Habitat (RGPH 3) риЬИё in 2002 by l'INSAE mentions 223,552 inhabitants including 106,097 men and 117,455 women. The annual population growth rate is 2.24%, i.e. an average of 3,584 births per year.

In 2013, the Recensement Gënëral de la Population et de l'Habitat (RGPH 4) риЫ1ё in 2015 by l'INSAE mentions 264,320 inhabitants including 126,016 men and 138,304 women. The annual population growth rate is 1.48%.

The population density in 1979 is ëvaluëe at 2561 hbts/Km^2 , in 1992 it is ëvaluëe at 3036 hbts/кт^2 , in 2002 this density is ëvaluëe at 4471 hbt/Km^2 and in 2013 this density has passedëe to 5083 hbts/Km^2 . Figure 40 provides a better illustration of this situation. The etiraeteristics of this population from the RGPH-4 are shown in Table II (page 84) and illustrated by Figure 43, which shows the Involution distribution of the town's population by district.

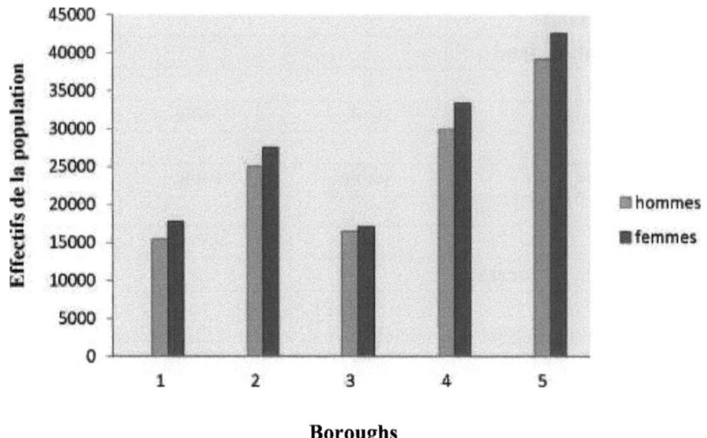

Figure 44: Rëpartition of the population by Arrondissement.

Source: INSAE/RGPH4 (2013)

The graph shows that boroughs 4 and 5 have the smallest populations, with 63,306 and 8,174 inhabitants respectively. In contrast, arrondissement 1, which is characterised by its old districts, has the smallest population with 3,316 inhabitants.

4.3.2 Demographic projections

Two main periods should be considered in the demographic growth of the city of Porto-Novo. Between 1961 and 1979, the population of Porto-Novo rose from around 61,000 to 133,163, with an average annual growth rate of just over 4%.

During the intercensal period (1979-1992), the population rose from 133,168 to 179,138, with an average growth rate of 2.3%, i.e. an average of 3,584 new inhabitants per year. This fall in the growth rate is essentially due to the decline in the migratory contribution. The average natural growth rate, on the other hand, rose slightly from 2.9% to 3.4%.

In 2002, the population was 223,552 and 264,320 (RGPH4, 2013) in 2013 with a growth rate of 1.48%.

By 2032, the population will have risen to 349,429, with a density of around 6,720 inhabitants/km^2 (figure 44).

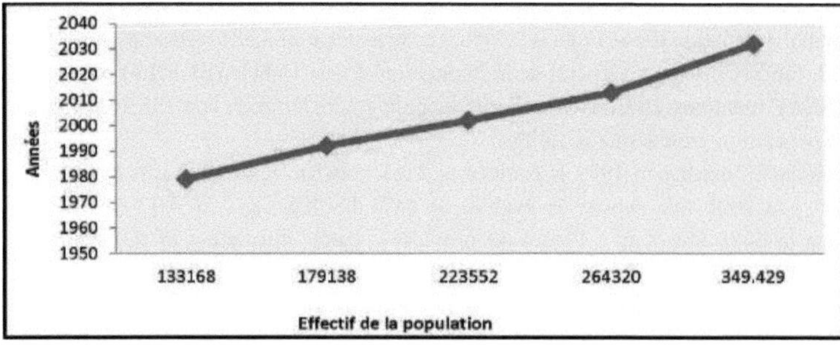

Figure 45: Population trends in Porto-Novo from 1950 to 2020 and projection to 2032

Source: INSAE/RGPH4 (2013)

This graph clearly shows the gradual evolution of Porto-Novo's demography. This population growth is having an impact on land use in the municipality^.

Following the example of the demographic revolution observed in Porto-Novo, the surrounding communes are following practically the same trends. Figures 45 to 54 illustrate the population situation in Akpro-Misserite, Avrankou, Adjarra, Agudguds and Seme-Podji respectively.

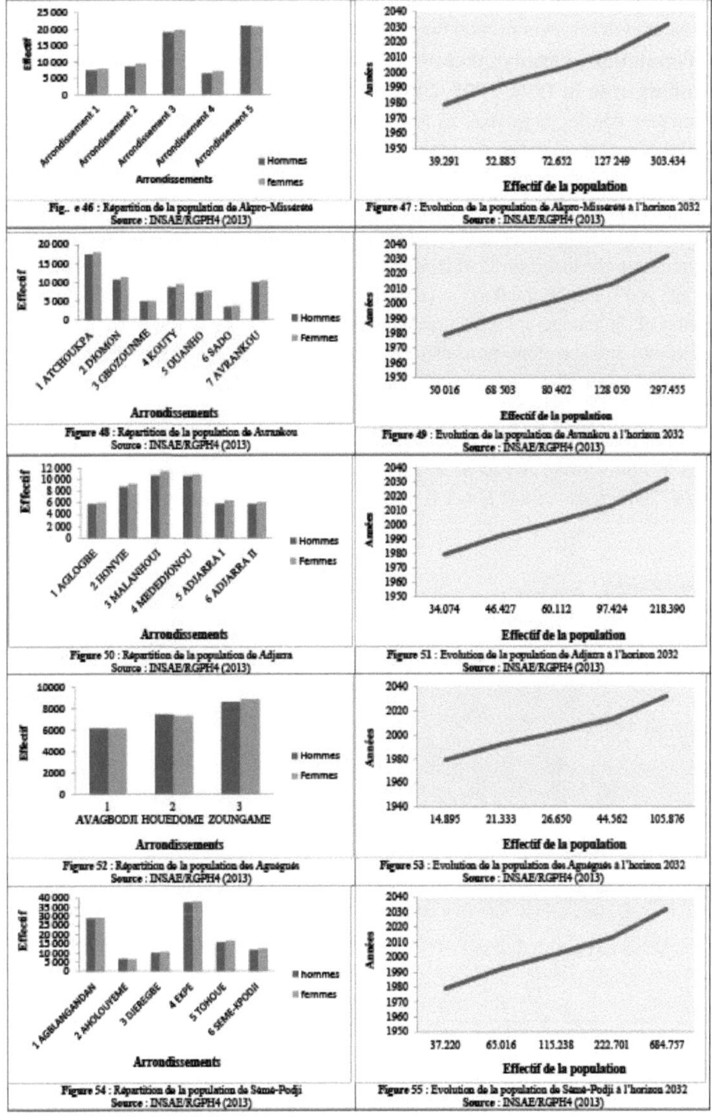

Plate 1: Population evolu... on in riverside municipalities

Analysis of all these graphs leads to the understanding that Гаддlотёгайоп urban Porto-Novo, surrounding communes included and even the one spёcialement lacustre, has experienced a sustained evolution since the 1950s. According to estimates, the communes of Porto-Novo, Лкрго-М188ёгёlё, Avrankou, Adjarra, Адиёдиёз and Sёmё-Podji will have 349,429; 303,434; 297,455; 218,390; 105,876 and 684,757 inhabitants respectively by 2032. This population growth has an impact on the occupation of the municipal territory^.

4.3.3 Population density trends in the commune of Porto-Novo and the surrounding area in 1979, 1992, 2002, 2013 and 2032

The population density maps (figs 55 to 59 below) clearly illustrate the progressive evolution of population growth in the commune of Porto-Novo and the surrounding area. By analysing this evolution with that of the agglomerations, it can be seen that the surface area of built-up areas has increased considerably, as has the population, to the detriment of natural formations.

A cross-sectional analysis of the figures shows that the city of Porto-Novo plays a very important role in the distribution of these observed densities. In 1979, this city had an average density of 1653 inhabitants/km^2 , with a population of over 130,000. Looking at the distribution of the orange-coloured areas, it can be seen that the Porto-Novo agglomeration is more colourful and therefore more densely populated. This trend is being maintained, with a progression and spread towards the other surrounding communes, which have also seen their population grow considerably.

In the light of these findings, the density projection for 2032 also shows an upward trend, with the cities of Porto-Novo, Лкрго-М188ёгёlё and Sёmё-Podji leading the way, with densities reaching 6720 hbts/km^2 , 3841 hbts/km^2 and 2739 hbts/km^2 respectively.

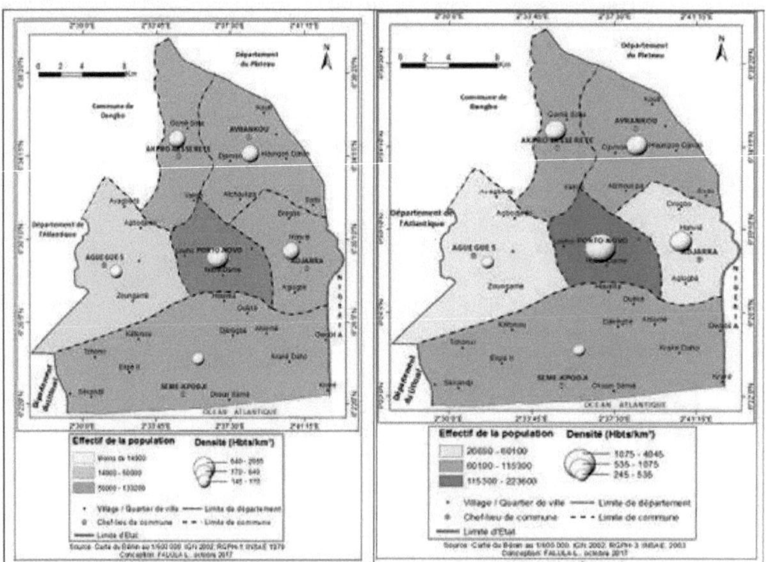

<u>Figure 56: Population density maps in 1979</u>
Figure 57: Population density maps in 2002

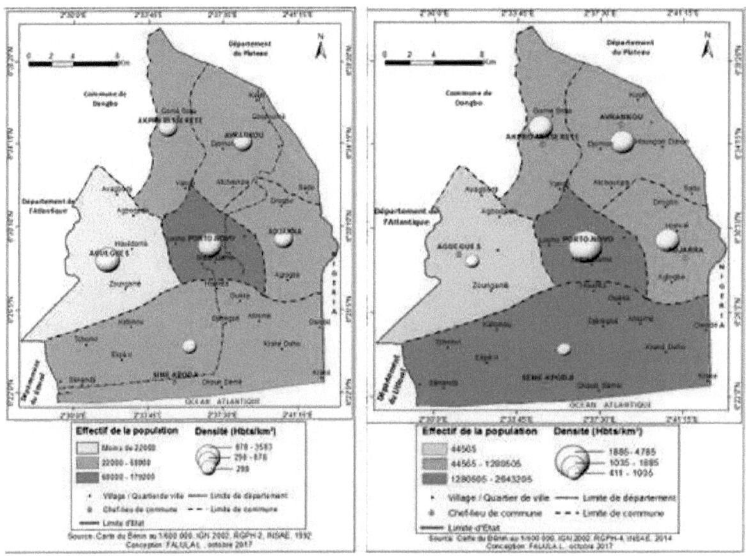

Figure 58: Population density maps in 1992
Figure 59: Population density maps in 2013

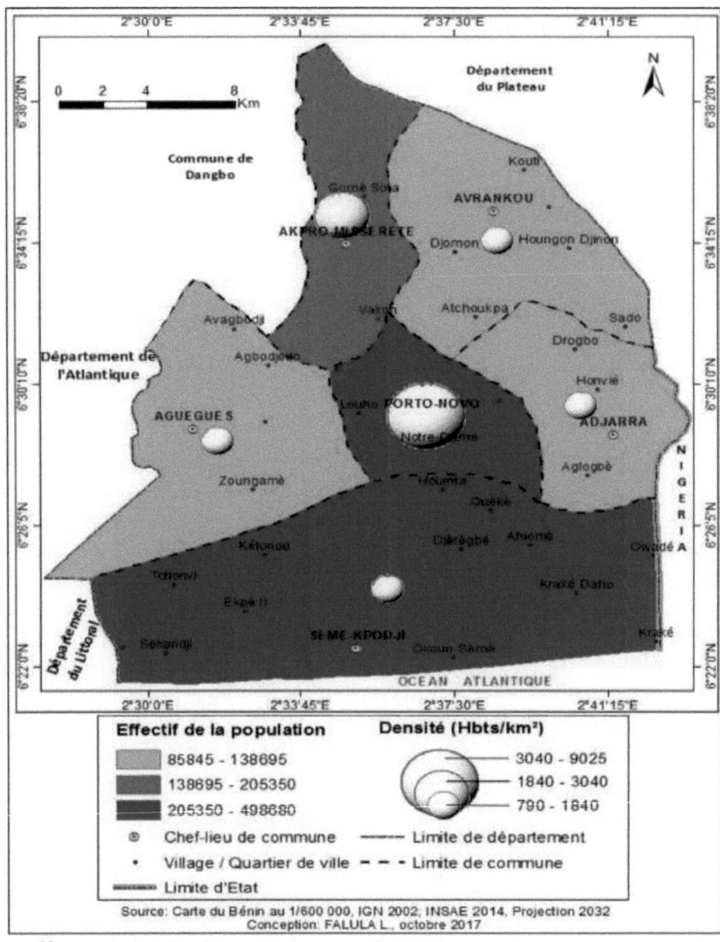

Figure 60: Population density projection maps for 2032

Correlation between land use units and population density

Table XXII shows the correlation coefficients calculated between the UOS.

Table 22: Correlation between UOS

Land use units	Correlation coefficient	Probability
Agglomëration	0.993	**0.001**
Culturespielieres	0.543	0.457
Honey palm crops	-0.881	0.119
sacred_forests	-0.999	**0.001**
Marecages	0.656	0.344
Cultivation a plantation	-0.855	0.144
Lake	0.638	0.362
Plantation	-0.715	0.285
Dënudë soil	0.523	0.477

Reading Table XXII, we note that the аддlотёгайопз present a significant positive correlation with dëmography, which means that the area of agglomërations increasedë with dëmography. The correlation of dëmography with sacred forests is also significant but negative, meaning that the area of sacred forests has significantly rëgressë from 1979 to 2012 with 1 population increase in the same përiod.

The other correlation coefficient values are not significant. However, we note a regression in the areas of most of the ипкёз of occupation with the dëmographic growth observed from 1979 to 2012 exceptë for marecages, water bodies and dënudës soils.

4.4. Discussion

This study provided evidence of the dynamics of change in the commune of Porto-Novo and its surroundings over the period 1972 to 1992, 1992 to 2012 and 2012 to 2032. The ëtude shows that areas of natural vdgdtation, i.e. sacred forest relics, are disappearing to the detriment of settlements, plantations and various crops. In other words, it is the disproportionate expansion of the Commune of Porto-Novo and its surroundings that is controlling the evolution of the landscape in the surrounding area. The driving force behind this expansion is essentially population growth.

According to INSAE (2014), the population density of the commune of Porto-Novo and its neighbouring communes shows the same pattern of development, i.e. essentially exponential growth. In 1979, the population density of Porto-Novo was estimated at 2561 inhabitants/km^2 , in 1992 it was estimated at 3,036 inhabitants/km^2 , in 2002 it was estimated at 4,471 inhabitants/km^2 and in 2013 it had risen to 5083 inhabitants/km^2 In 2032, the population is estimated at around 349,429 inhabitants, with a density of around 6720 inhabitants/km .^{2}

The ever-increasing population puts considerable pressure on natural resources and, in many cases, leads to their degradation and depletion. In other words, the study shows that natural space disappears over time and cultivated areas increase with population growth.

These results confirm those of Fangnon *et al* (2013) in the department of Couffo in Benin, those of Tente *et al* (2011) in the communes of Glazoud and Dassa-Zoumd in central Benin and those of Djohy *et al* (2016) in the commune of Sinende in northern Benin. Their results show the pressure exerted by the population on the vegetation cover, i.e. demographic pressure, which is reflected in the intensity of human activities in general and agricultural activities in particular, is the main cause of the degradation of the vegetation cover. Similarly, Aminata (2006) has shown in her work that the rapid urbanisation of the Dakar region has altered almost all the natural areas, including ponds and natural vegetation. MOUSSA (2006) has shown that human activity is responsible for the degradation of the environment. OUSSEINI (1994) has shown that human concentration is a function of the availability of natural resources and that demographic dynamics are a real threat to them.

The work of Taibou Ba and Dieynaba Seck. (2012) in Sdndgal indicate a general trend towards the artificialization of the matdrialised zone through the gradual extension of agricultural areas, the salinization of land, the loss of natural vegetation, especially mangroves which play a decisive role, not to mention the increase in population. The same observations have been made in the commune of Djougou in Benin (Leroux L., 2012), with the expansion of areas devoted to cultivation and urban areas taking place at the expense of areas of dense natural vegetation (dense open forest, gallery forest, classeg forest), which are undergoing a sharp decline. This confirms the results of work carried out in northern Côte d'Ivoire by Tanina *et al*, (2013), and in the city of Niamey and its përiphërie by Sanda Gonda H. (2010). This could be explained by the fact that the use of these areas makes it possible to bënëflcier new fertile land and thus increase agricultural production. In other words, there is increasing pressure on natural resources, driven in particular by the rapid cultivation of natural vëgëtation areas in order to meet the food needs of a growing population (Leroux L., 2012; Tanina D. et *al.*, 2013; Sarr, 2009).

It is therefore necessary to be able to estimate spatio-temporal variations in land use. In this sense, regular monitoring of changes in land use, as was done in this study, shows that land use is a fundamental variable in environmental management and in understanding how it functions in relation to climatic factors.

In this way, the dynamics of land use depend on the demographic evolution of the population and its density.

Alternative measures must therefore be taken to limit environmental degradation and ensure the survival of future generations. Rational management of space is therefore essential.

Partial conclusion

In view of the results obtained, we have found that the land cover has shown very significant changes over the different dates: 1972, 1992, 2012 and 2032. We can see that the most widespread land use units are palm crops and fallows, crops and fallows, plantation crops   , wetland   formations, water bodies and agglomerations, which occupied more than 94.72% of the total area in 1972, 95.94% in 1992 and 97.29% in 2012. In 2032, these classes will occupy more than 96.96% of the total surface area.

Finally, a visual analysis of the various land use maps shows that the main changes are as follows:

a)   the expansion of built-up areas or conurbations;

b)   the decline and even disappearance of sacred forest relics;

c)   the growth of palm crops and fallow land ;

d)   growing crops and fallow land

e)   the growth of plantation crops

The classes that appear to be the most stable over the various dates are: bare ground, water and wetland formations.

During the different periods 1972-1992, 1992-2012 and 20122032, the areas of natural vegetation (relics of sacred forests) disappeared to the detriment of agricultural areas (crops and ptclieres, palm crops and ptclieres, plantation crops) and built-up areas (agglomerations). These different results allow us to conclude that hypothesis 1, i.e. that land use undergoes an essentially progressive evolution in the commune of Porto-Novo and its surroundings, is vëriflëe. In the following chapter, we will determine the influence of building density and green spaces on the variation in UHI from 2001 to 2015.

CHAPTER V: BUILDING DENSITY AND THE ROLE AND INFLUENCE OF GREEN SPACES ON ICU

The results presented in the preceding chapters demonstrate the links between land use dynamics and urban population growth. The density of built-up areas should also follow the same pattern. In this chapter, therefore, the aim is to characterise building density and the role and influence of green spaces on UHI.

5.1. Criteria reflecting building density

Spatial criteria are defined to characterise the types of building density in the study area. Thus, the different types of housing often found in the various districts/villages of our study area, with different faults and administrative status, located in different geographical areas, can be classified according to three simple degrees of built density (low, medium and high density), measured at the scale of the plot or block (Ringenbach, 2004). Table XXIII shows the physical characteristics of the different types of housing in the study area and their building densities, estimated on the basis of floor plans.

Table 23: Physical characteristics of the different types of dwellings in the study area and their estimated built density based on floor plans

Degree of built	Types of dwellings and housing estates commonly found in the various neighbourhoods/villages in the study area	Main spatial characteristics specific to some of the group's residential developments
Low building density	1. Housing estates 2. Suburban housing estates 3. Other housing estates and cooperatives 4. Semi-detached or multi-storey housing, medium-sized towns and villages 5. Medium-sized and small towns : 6. Housing or individual houses of a particular shape 7. Housing or detached houses along major roads 8. Illegal housing (shanty towns), on the outskirts of large districts	1. Detached houses on large plots with gardens, built on very large open plots or detached houses on their own plot, not adjoining adjoining buildings, large outdoor spaces, semi-private roads. 2. Subdivisions with detached or semi-detached single-family dwellings, built around large driveways. 3. Relatively low buildings, large plot sizes. 4. 2-storey building, individual or shared accesses, small footprint. 5. Plots with long trapezoidal shapes and narrow frontages, making it impossible to build a second row. 6. The building is set back from the carriageway, leaving a large space for easements. 7. Very small plots, tightly woven fabric, narrow lanes. 8. Group housing, not attached to each other, set back from the roads, around large empty spaces
Average densities	1. Resettlement areas and informal housing estates. 2. Large recent housing estates and the outskirts of the city. 3. Collective housing. 4. Town houses in or near major urban centres. 5. High-rise housing complexes along major urban routes.	1. Houses built alongside narrow lanes, high extensions, long blocks. 2. Vertical urban development, disappearance of the plot grid, discontinuous layout of large buildings, large open spaces, with carriageways making up the majority of the space left free. 3. Group of dwellings, discontinuous layout, set back from roads. 4. Attached, single-storey houses along the lanes, with integrated commercial activities. 5. Single-family dwellings in continuous order on the street.
High	1. New promotional housing developments and social participative housing, close to major urban centres and towns	1. Recent collective housing buildings aligned with the streets bordering the block, farmhouses or semi-farmhouses. Traditional houses, introverted around open or partially converted patios, narrow winding lanes, sometimes covered, cul-de-sacs providing

2. Major colonial complexes	access to the dwellings, a few main roads, also pedestrianised, where public activities are concentrated, tight, dense urban fabric. 2. Colonial-style urban fabric, with a proliferation of semi-detached buildings; these blocks of flats are uniform in height (R+5 + sometimes attic).

Source: (SAS.Planet) and views of the Adriatic (Google Earth)

The питёпзайоп of the buildings resulted in polygons whose area is а1зётеп1 сагас1ёпзёе and exprimëe. Figure 60 shows an overview of the питёпзайоп window under 1 Google Earth application.

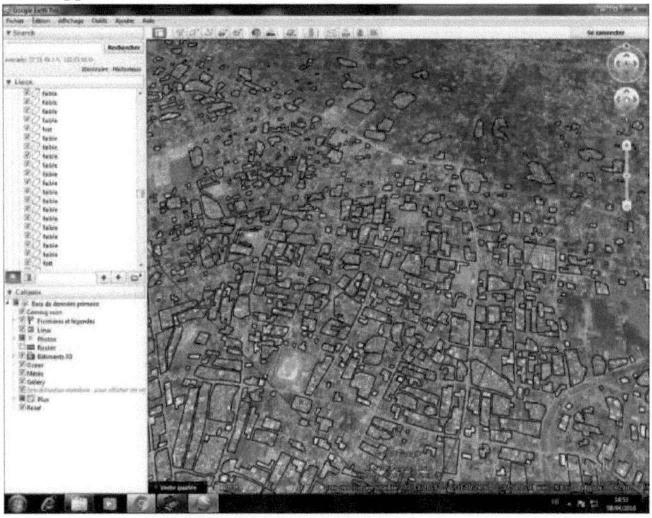

Figure 61: Numbering of buildings in Google Earth Pro

Source: Falolou, April 2018

We note that there are different possibilities for building individual and collective housing, the type of housing and the number of storeys not always determining a real "high density"; certain forms of individual housing can be dense. It should also be noted that the built density of residential complexes made up of informal housing or housing estates, townhouses, traditional housing (...) catëgorisës of individual types, having an average number of levels Нткё 2 or 3, dë exceed the densities of most apartment blocks, including those ctiraclenstint 1 urbanisme vertical de la përiode postcoloniale (Centre de Porto- Novo) et notamment recente, constit^ par 1 implantation discontinue de grands immeubles (Hotels, administration) avec espaces ouverts importants.

5.2 Mapping of digitised buildings

The питёпзайоп of the existing buildings in 1 emprise of our dktude area made it possible to obtain the maps below. These maps represent the питёпзёз buildings in the commune of Porto-Novo and its surroundings. In other words, there is a concentration of buildings of varying densities.

We observe a Ьё1ёгодёпёкё of habitats, relative to the measure of 1 land use intensity of each commune in the study area.

Thus, based on the data from this study, we were able to identify 3 classes of building density

in our arid environment (Table N°):
-    low density: < 14 lots / Km^2 ;
-    average density: between [15 lots / Km^2 - 39 lots / Km $]^2$
-    high density: > 40 lots / Km^2

On reading the various maps, it should be noted that all the communes apart from Aguegue have all 3 density classes, which means that these communes are undergoing a major spatial transformation. Precisely, this density has been mapped according to each commune and their more detailed analysis has made it possible to understand the evolving trends from a local point of view.

o **Porto-Novo**

Analysis of the density of Porto-Novo reveals that the commune of Porto-Novo is densëment bati (Figure 61). However, there are a few isokic pockets that are an exception. These correspond to the dëpressionary surfaces located around Djassin (Zounvi depression), Louho and Dowa (Boë depression), Agbokou (Donoukin depression) and around the lagoon bank at the southern exit from Porto-Novo.

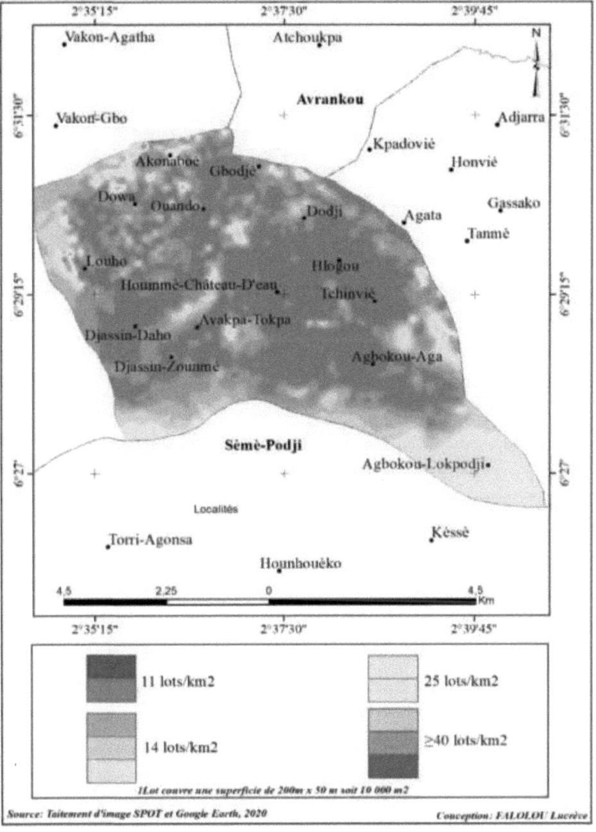

Figure 62: Building site in Porto-Novo

In addition, it should be noted that this high density оѣегуёе seems to impact the рёпрѣёпе

North of this capital city, particularly towards the communes of Avrankou, Adjarra and Akpro- M188ёrёlё.

o **Avrankou**

The commune of Avrankou has two major areas of high density: one in the south around Atchoukpa and the second in the north-east around the Gbëtchou, Dangbodji and Ahigon localities (Figure 62).

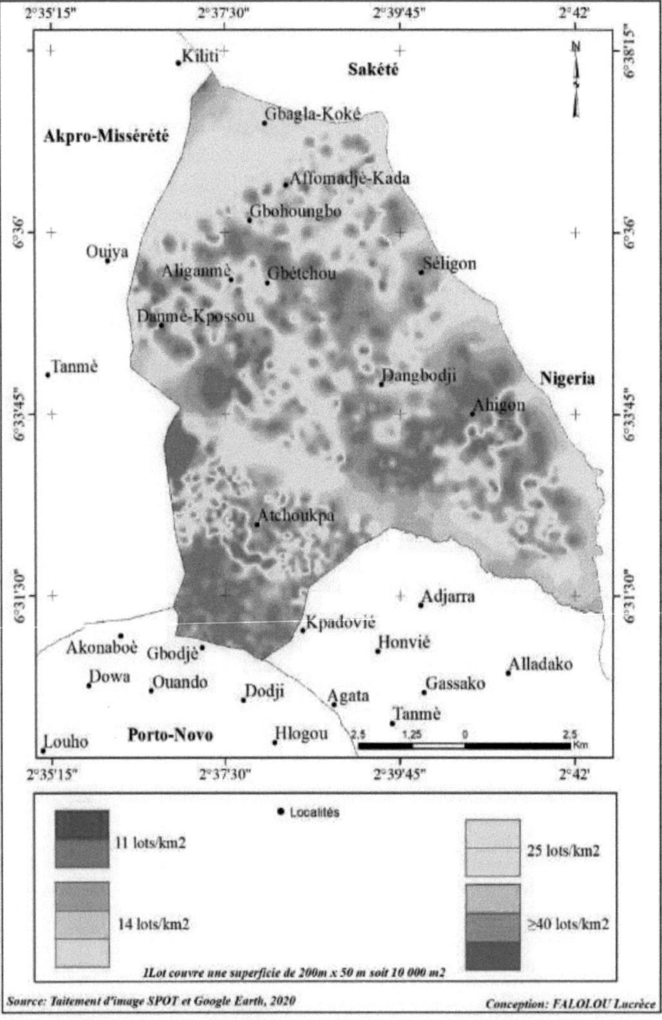

Figure 63: Building density at Avrankou

The area covered by high-density beaches occupies around 50% of the total surface area, while medium-density areas cover 35% of the total surface area and low-density areas 15%.

o **Adjarra**

The commune of Adjarra, unlike that of Avrankou, shares a large part of its boundary with the

other communes.
West with the capital city (Figure 63).

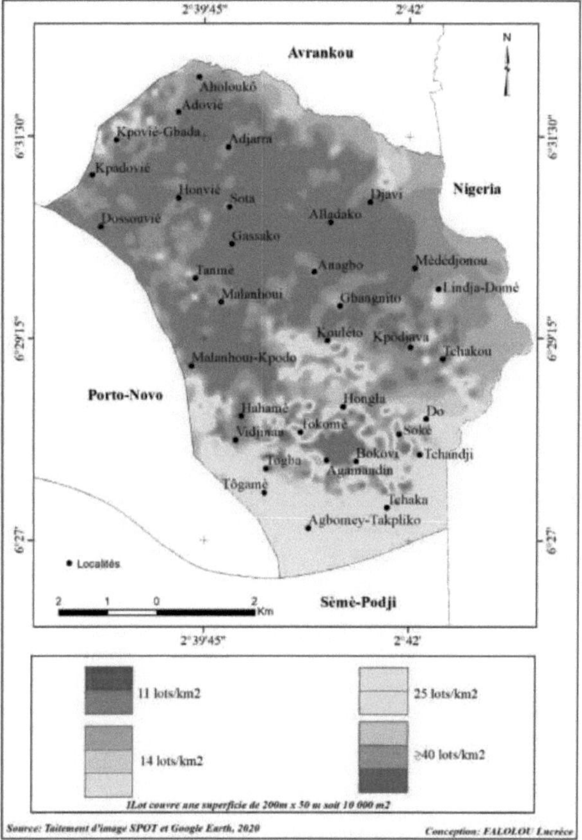

Figure 64: Building density in Adjarra

The trend observed in Porto-Novo seems to extend naturally towards Adjarra, where more than half of the area is high density. Only the southern part of this commune shows a low building density trend.

o **Misserete case**

This commune, which stretches from south to north, also borders on Porto-Novo (Figure 64).

103

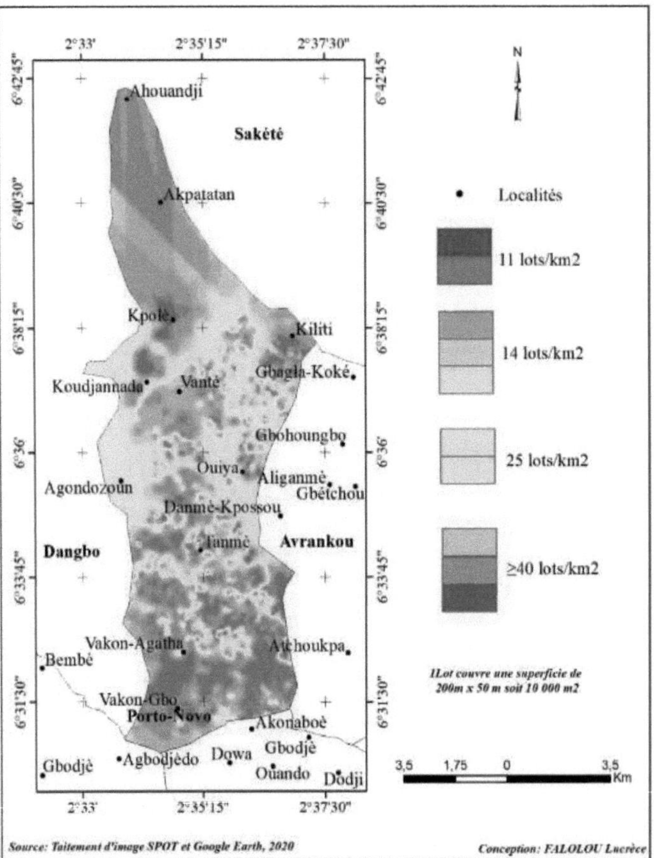

Figure 65: Building density a Akpro-Missdrdtd

The trend observed between Adjarra and Porto-Novo seems to be repeated here, since the high density observed in Porto-Novo extends over much of the southern part of this commune. If the trend continues, the last northern portion (a third of the commune's surface area) will be more densely populated.

is in danger of being caught up. It is important to note, however, that there are a few pockets of low density, particularly in the centre.

o Seme-Podji

The commune of Seme-Podji, located on the coastal plain, has a well-targeted spatial layout given the morphological nature of the site (Figure 65).

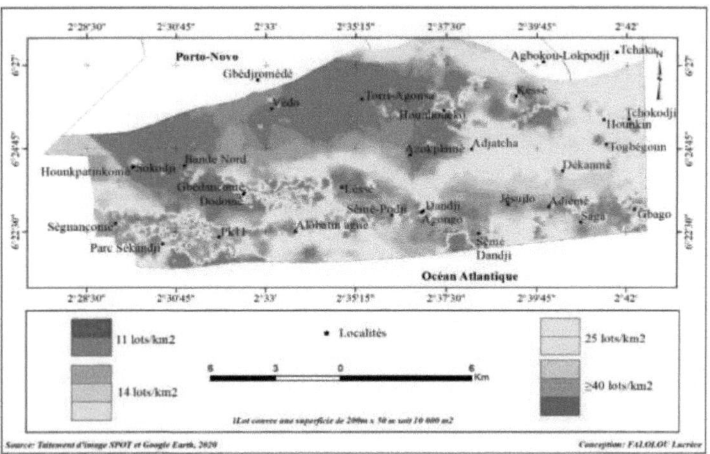

Figure 66: Building density in Seme-Podji

In fact, by observing the spatial distribution of the areas of colour that provide information on density, it can be seen that the settlements are located on coastal strips bordered by marshy depressions. The densest portion is to the south and is more or less discontinuous (Sekandji, pk11, Podji saga,). The least dense portion can be seen to the north-west of figure 65 (Tori-Agonsa).

○ **Case of Aguegues**

The Agudguds commune has a low density of houses along the lagoon, with the distinctive feature of houses on stilts (Figure 66).

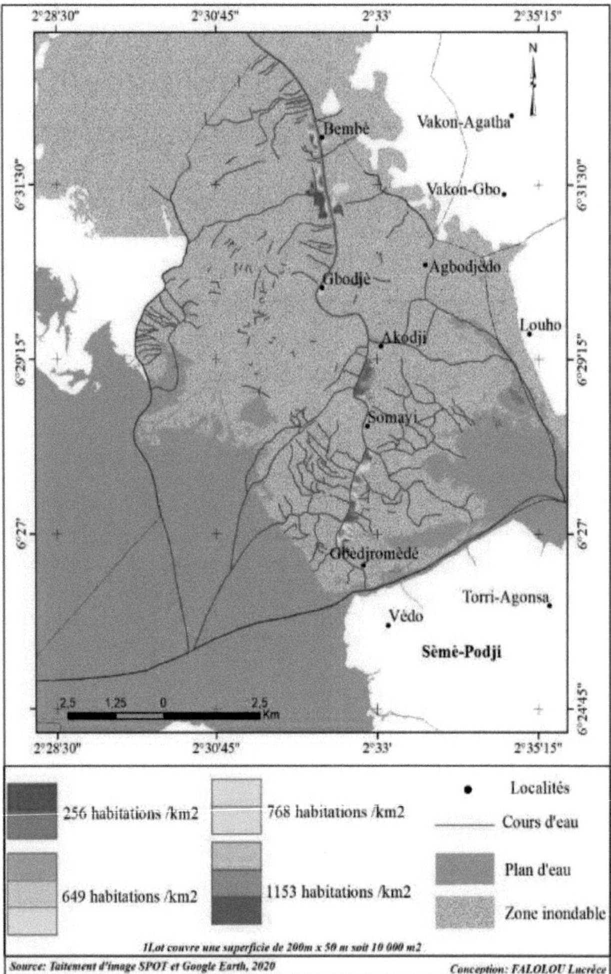

Figure 67: Building density at Лдиёдиёз

If you look at the figure above, you can see areas of varying density following a linear distribution. These areas are particularly dense in the south of the municipality.

In general terms, our study area is densely packed, with a huddle of buildings in places that does not encourage easy wind circulation, and a few high-rise buildings that create urban canions (Figure 68). High building densities are spatially distributed to the south on the one hand and from the centre to the north on the other. These results confirm that the density of the study area is undergoing a gradual evolution in line with the size of its population.

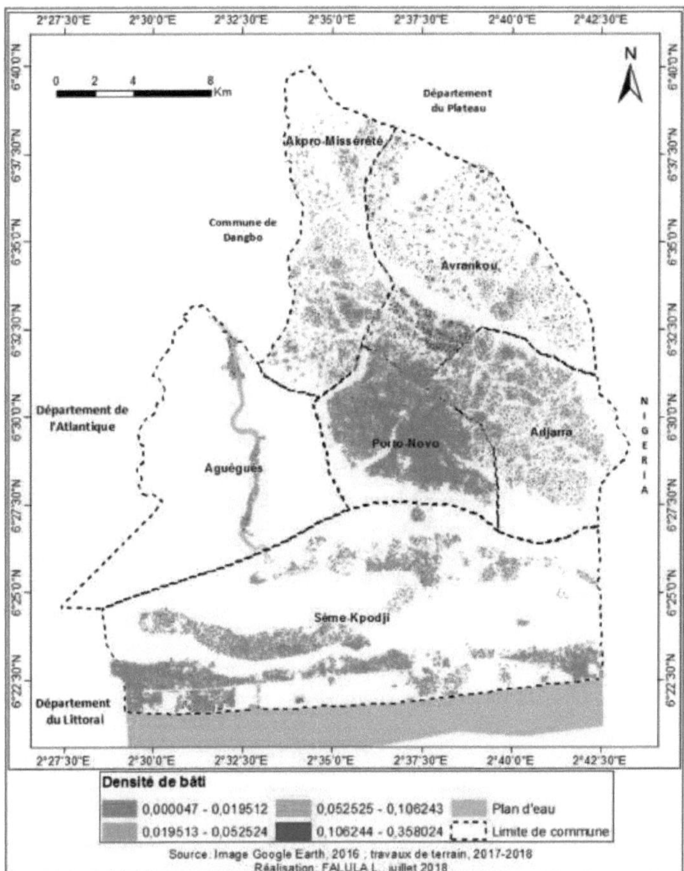

Figure 68: Building sprawl in Porto-Novo and the surrounding area

5.3 Characterisation of green spaces, their role and influence on UHI

5.3.1 Mechanisms for vegetation action on urban heat pilots

To counter some of the dësagrëments and inconveniences of today's urban context, the presence of yëдёlайопо plays a non-iiegligeaNe, even indispensable role. They have multiple bënëfices. Indeed, the many benefits of plant-based vëgëtals no longer need to be proven; however, it is still relevant to highlight their main contributions at the level of the environment and populations.

5.3.2 Contribution of the Porto-Novo JPN to the purification and improvement of the quality of Pair

The total aerial biomass of the botanical garden conservatory (Jardin des Plantes et de la Nature (JPN)) in the city of Porto-Novo is estimated at 92771 t/ha and that of site 2 (the storage site) is estimated at 67470 t/ha. These aërienne biomasses are the carbon reservoirs in the vëgëtation of the botanical garden. Table XXIV shows the carbon stock in the aërial biomass of the two botanical garden sites.

Table 24: Carbon stock in the aërial biomass of the two JPN sites.

	Site 1 (conservatory)	Site 2 (relaxation area)

| Total biomass (t/ha) | 92771 t/ha | 67470 t/ha |
| Carbon stored (t C/ha) | 463.9 t C/ha | 337.4 t C/ha |

Source: Fieldwork (December 2017)

From this table, we can say that the carbon stock in the aërienne biomass of the conservatory (site 1) of the JPN is higher than that of site 2 (area of dëtente). This can be explained by the fact that site 1 (conservatory) of the JPN is richer in plant species than site 2 (holding area). Indeed, site 1 contains more than 200 species of tropical and acclimatised vëgëtales, whereas site 2 contains only 167 species. Similarly, all the different vëgëtal species in site 2 are also found in site 1 (conservatory). Also, the rëpertoriës trees (137 in total) have an average height of 14.81 m with an average circumfërence of 1.38 m (appendices).

So, in general terms, the more trees there are, the higher the emumigasin carbon qmmtite, which easily leads to a reduction in greenhouse gases. Otherwise, these vëgëtals act as absorbers of gaseous pollutants (NOx, ozone, CO2, VOCs, etc.). ***Hence the justification for the chlorophyll and anti-pollution functions of green spaces.*** These functions are in fact provided by the sun's light energy and the chlorophyll in the leaves. By consuming carbon dioxide from the air, trees produce their own matter and release oxygen. This makes them veritable oases of coolness. In addition, trees cast shade on the ground, which reduces the potential heat that minëralisëes surfaces can absorb and helps to protect the population from direct sunlight (Vergriete and Labrecque, 2007). This shade reduces the amount of heat that buildings, roads and other structures can absorb and then release later (Gendron-Bouchard, 20 13), and provides people with places to take a cool break on hot summer days (Allain, 2004). Finally, trees and other tall plants act as natural shades, blocking and filtering some of the ultraviolet rays before they reach the ground and human activity (Vergriete and Labrecque, 2007).

The leaves also retain dust and other particles on their surface. In a planted street, there is four times less dust than in a non-planted street; this can represent almost 100 kilograms of dust fixed by each tree every year (Abbey, 1993). Leaves clean the ambient air by retaining micro-organisms mixed with the dust. They are also capable of absorbing and using certain gaseous pollutants (carbon monoxide, nitrogen dioxide) (Richard, 2006).

Trees in particular and vegetation in general play a role in attenuating noise, particularly in the case of large wooded screens (Tossou, 2007).

Through their transpiration, trees release water vapour into the atmosphere, thereby helping to correct the climate in our cities through their cooling action. They also have a "purifying" role, as they photosynthesise to absorb the carbon released by the combustion of motor fuels. They help to combat air pollution (lower temperatures, less dust) (Richard, 2006).

Plants produce evapotranspiration: they use heat as energy and release water vapour into the air, which helps to cool the ambient air (MAMROT, 2010). For example, "a mature tree can extract more than 450 litres of water from the soil and then release it into the air in the form of water vapour [...] a cooling effect equivalent to that of five air conditioners operating 20 hours a day" (Vergriete and Labrecque, 2007 :

8)   . By reflecting part of the sun's rays and reducing the absorption and emission of heat, plants help to mitigate the effects of urban heat islands (CRE- Montreal, 2007). This is an important issue, since UHIs are a growing public health problem that often affects the most socially and materially disadvantaged populations (Cavayas and Baudouin, 2008). These vëgëtaux thus play the role of thermal bënëfices. (CREMontreal, 2005, n.p.).

Similarly, these vëgëtals also act as solar screens in summer and protect against winds in winter (cold periods) (CREMontreal, 2005, n.p.).

Trees in the city help to make the urban environment more tempërë and mitigate the impacts of heat islands on populations by reducing tempërature variations in summer and winter, acting as ee rails to attenuate winds, protëgating ultraviolet and infrared rays and cooling ambient fair, (CREMontreal, 2005 and 2007 ; Gendron-Bouchard, 20 13; Yergriete and Labrecque, 2007).

In addition, Genin and Plantineau (1982) demonstrated that one hectare of forest releases the volume of oxygen necessary for a man to breathe for 25 years. In view of this result, we can conclude that the need to promote green spaces in urban areas is justified because of the increase in greenhouse gases and the proliferation of the UHI phënomëne.

As prëcëdently mentioned, green spaces fulfil various functions within a city. In addition, trees in particular promote and contribute to the formation of cooler zones in the urban environment that reduce the intensity of UHI phënomëne.

In short, it can be said that green spaces encourage and contribute to the formation of cooler zones in the urban environment, which could reduce the intensity of the UHI phënomëne.

These results confirm the hypothesis that building density and green spaces influence the variation in UHI in the commune of Porto-Novo and the surrounding area.

5.4 Discussion

Building density

The results obtained on the demolition of buildings in the commune of Porto-Novo and the surrounding area are in line with those of a number of authors who have used various methods and models in their research.

Thierno Aw (2010) has shown that the essence of the city lies in the variety and density of the exchanges it offers its citizens. These characteristics depend on the quality of the urban structure, which in turn is conditioned by a number of structural elements marked by causal links. He also argues that the main theoretical approaches that seek to explain, for example, the interactions between transport and land use are based mainly on economic, technical and sociological theories. These theoretical considerations relate to three areas in particular:

- The location choices made by economic agents (households, businesses, institutions) and their impact on the transport system,

- transport system choices (institutions) and their consequences for spatial practices (personal mobility: choice of mode, motives and types of travel),

- the choices made by individuals to engage in social interaction, in close connection with location choices (places of co-presence) and mobility trade-offs (the practice of an activity being conducive to social interaction).

These theories presented by this author corroborate the results obtained in this thesis.

On a completely different note, some pertinent questions have arisen concerning the concerns relating to the objects to be studied; at what scale should they be studied and what measure should be used to quantify them. On this subject, Girard M. (2017) provided some answers in one of his articles entitled: "Organisation spatiale et densitds urbaines: une application a I'agglomdration du Grand Dijon". For this author, an urban density can refer to both a 'container' density (buildings) and a 'content' density (population) (Fouchier 1997).

Thus, depending on the purpose of the study, it is possible to distinguish quantitative areas (more or less densely populated or urbanised) and to characterise them (what population makes up an area, what urban forms dominate an area). The approach used in this thesis to calculate building density is similar to that of this author if we focus on the principle of the 'container' (building).

The author also draws attention to the spatial resolution of the data to be mapped and analysed. For the purposes of this thesis, the results obtained at cadastral and regional scales show that the higher the density, the greater the spatial resolution of the data used.

Analysis of carbon stock in aerial biomass

The carbon stock in the Adriatic biomass of the botanical garden conservatory is 463.9 t C/ha and that of site 2 (ddtente) is 337.4 t C/ha, giving a total of 801.3 t C/ha. So the JPN alone has the capacity to absorb 801.206 t C/ha. This means that if there were several green spaces like the JPN in Porto-Novo and the surrounding area, they would help to reduce pollution, particularly greenhouse gases such as carbon dioxide (CO_2). This result is higher than the range given by Palm et al (2000), between 40 and 60 t C /ha. It is also higher than the value obtained by Peltier et al. (2007), who found a carbon stock of 5.046 t C/ha in the aërienne biomass of a karite stand in North Cameroon. It is also higher than the ranges given by Albrecht and Kandji (2003) and Bello et al. (2017) who estimate it at between 7 and 25 tC/ha in an agroforestry system; and Medlyn et al., (2013), at between 24.42 ± 6.98 t C /ha in karite stands and пёrё de Bembёrёkё. Similarly, our results (463.9 t C/ha; 337.4 t C/ha) are significantly higher than the 20.08 tC/ha obtained by Odiwe et al. (2012) in a Tectona grandis plantation in Nigёria.

The carbon value at the two sites (801.3 t C/ha) in this study is high (ёlevёe) due to the predominance of large-diameter woody species and their density. This value is lower than that obtained by the IPCC (2001) for the French forest, which is 1,220 t C/ha. This is why Brown & Pearson (2005) believe that carbon stock assessment should be limited to woody species. Other authors (Valentini, 2007; Sai'dou et al., 2012), have confirmed "these remarks" by explaining that the herbaceous flora has a low contribution to the total volume of carbon.

According to the IPCC (2001), vegetation plays a fundamental role in limiting climate change and mitigating urban heat islands, since it acts as a carbon sink.

In   carbon carbon plays an   important role in   changes   climate change . It has two main roles: (1) as a greenhouse gas when it is in the form of CO_2 and (2) as a climate change mitigator when it is stored. The reduction of carbon dioxide by trees is an important ёcosystёmique activity.

Following the same line of thought, Xavier K. et al (2019) determined the organic carbon stock in Acacia auriculiformis plantations in classified forests (Ouedo and Pahou) in order to understand their contribution to climate change mitigation. Their results showed that the carbon stock is higher in the trunk: 78.17% than in the branches: 19.50% and the leaves: 2.33%. Soil carbon sёquestration potential ranged from 31.46% to 70.99% depending on soil type. The carbon stock is significantly higher in the aerial biomass than in the root biomass. With this in mind, Montagnini and Nair (2004), have shown that carbon capture efforts in tropical environments are concentrated on aerial biomass and not in soils.

Also according to Xavier K. et al (2019), the highest carbon stocks were obtained in the 2.5-year-old plantation on hydromorphic soil in Pahou (106.23 ± 38.39 tC/ha) compared with 77.77 ± 5.47 tC/ha for the 2-year-old plantation on ferralitic soil in Ouedo. Moreover, the latter demonstrated that there is a strongly negative and significant correlation (r = - 0.931, P < 0.001) between tempёrature and carbon stock. The high tempёratures observed are the result of low carbon sёquestration and vice versa. Their results corroborate those of Boulmane et al. (2013); Bello et al. (2017) and Diatta et al. (2016) who obtained similar results in Morocco in the green oak forest, in Bёnin on cashew plantation and in Sdndgal in the South of the Groundnut Basin.

Carbon dioxide is a greenhouse gas that is primarily responsible for climate change (Richard,

2006) and urban heat islands. The Intergovernmental Panel on Climate Change (IPCC) estimates that global CO2 emissions should not exceed 10 billion tonnes per year by 2050 if we are to avoid environmental repercussions that could prove particularly catastrophic for developing countries (Richard, 2006).

Carbon dioxide is a gas that occurs naturally in the atmosphere. As they grow, plants absorb carbon dioxide, which combines with water to create simple sugars. When a plant dies, burns or is eaten by an animal, the carbon is released back into the atmosphere. The world's forests and forest soils store over a trillion tonnes of carbon, twice the volume present in the atmosphere (Richard, 2006). The destruction of forests also injects nearly six billion tonnes of carbon dioxide into the atmosphere every year (Richard, 2006). Vegetation absorbs and stores carbon dioxide (which photosynthesis converts into carbon), acting as a 'carbon sink'. But conversely, when these plants are destroyed or overexploited and burnt down, they can become sources of this same carbon dioxide.

Partial conclusion

In general terms, our study area is characterised by overcrowding, with a tight grouping of buildings in places that does not encourage easy wind circulation, and also by a number of storey buildings that create urban canions. High building densities are spatially distributed to the south on the one hand and from the centre to the north on the other. These results confirm that the density of the study area is undergoing a gradual evolution in line with the size of its population. There appears to be a relationship between the variation in UHIs and the observed building density. Areas of high density should coincide with areas of high density. Similarly, it has been found that the presence of green spaces creates a cool zone and thus lowers the temperature of the environment; hence the reduction in the intensity of the UHI.

CHAPTER VI : Characterisation of UHIs, their impact on the environment and approach to sustainable solutions

In this chapter, observation of the hottest months has been used to characterise UHIs and analyse their impact on the environment and on human health. Finally, this chapter will propose solutions for the sustainable regulation of UHIs.

6.1. Determining the hottest months

Most studies on UHI focus on summer periods solely because of the climate of these regions. But in the present study, due to the alternation of seasons, we have deliberately chosen to focus our attention on the hottest months consecutii's of l'annëe based on l'ASECNA data over the period 1985 to 2015 (Figure 67).

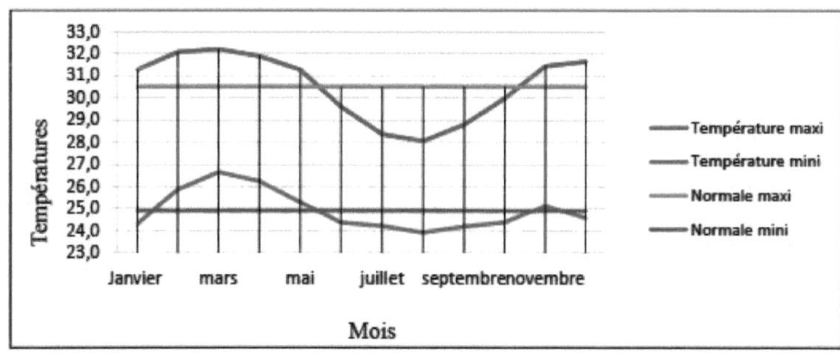

Figure 69: Average monthly tempëratures over the përiod 1985-2015 in Porto- Novo and the surrounding area.

Source: ASECNA, Meteo Bdnin

Figure 67 shows that maximum and minimum temperatures have followed the same pattern over the period 1985 - 2015. The parts of the curve (the maximums or minimums) at the bottom of the normal represent the least warm months and those at the top of the normal represent the warmest months. For maximum temperatures, the warmest months of the year are: January, February, March, April, May, November and December, with an average value of 30.5°C. On the other hand, the hottest months for minimum temperatures are February, March, April, May and November, with an average value of 25°C.

In southern Benin, according to ASECNA data, the long drying season runs from December to March and sometimes from December to mid-April, while the short drying season runs from mid-August to mid-September. This study concentrates on the long drying season of each year studied over the period 2001 to 2015 in order to obtain better satellite images and to better study heat islands over a long period of the year and avoid the overlap of two years, for example December 2000 to April 2001. Since the study of heat islands focuses mainly on heat extremes, we have chosen the hottest months on the basis of the maximum temperature curve. Figure 67 shows that the months of January, February, March, April and May were the hottest. The month of May, which is at the lower end of the temperature curve and also forms part of the rainy season, was eliminated from the analysis in order to reduce cloud excess.

For the purposes of this study, the months of January, January, March and April have been selected.

6.2. Drawing up heat maps of the study area

Map sëries are rëaH3ëe3 to allow a clear visualisation of the thermal dynamics of Porto-Novo and its surroundings. Plate 2, presents the distribution of tempëratures over the fifteen (15) dates sëlectnëes, during the months of January, fëvrier, March and April.

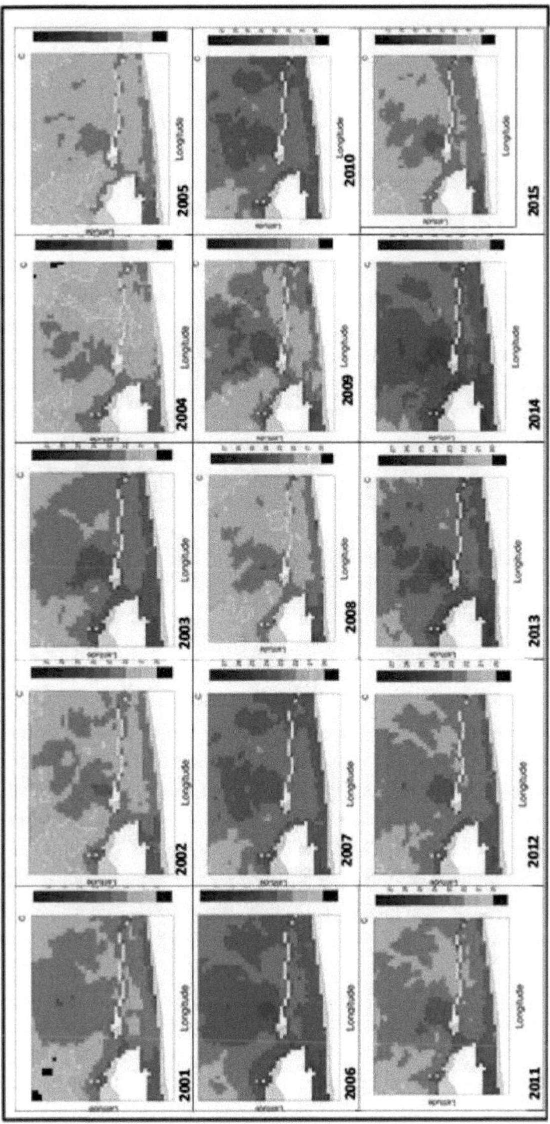

Figure 2: Temperature distribution from 2001 to 2015 in the dry season

Source: MODIS images; Land Surface Temperature

The зёпез maps of the 1етрёгаШге surface from MODIS images of the 15 аппёез (2001 - 2015) show, on the one hand, 1 progressive installation of the warm air mass and, on the other hand, an intensification of 1 ICU with consequential ёcarts between the city centre and the përiphërie. The images depict different thermal surfaces of the ёШШё landscape using a graduated colour scale.

Thus, the classification of these figures into colour ranges is identical for each of the dates selected, without distinction between the ttnnees. The colours represent different ground temperature values, with the emphasis on the warmest surfaces in the study area. This scale, of graduated colours, evolves from orange-red to dark red for warm surfaces, and from blue-green to yellow for cool surfaces. Consequently, heat islands are found in dark-coloured areas (from orange to reddish-ochre). These dark-coloured zones appear to induce overpowering heat for the human body at night, with temperatures above 22°C.

In addition, the use of these colours on satellite images, particularly in our study area, reveals the thermal characteristics of different types of ground use typical of urban areas, such as buildings, paving, tarmac, asphalt, impermeable surfaces and vegetation.

Orange, red to dark red are concentrated on building roofs, impermeable surfaces, asphalt and the dark colour cttraclerise the interior of cities; which ranks them in preт!ёre position of the warmest surfaces. Blue and green cover several surfaces still inside cities, mainly thanks to the shade caused by certain buildings as well as surfaces covered with vëgëtation and wet surfaces. The yellow colour indicates the temperature between the coolest and hottest areas, i.e. the yellow colour determines the spatial thermal average for each image series.

Looking at these map series, we can see that the maximum temperatures increase in intensity and surface area (orange, red to ochre red) over the years. As a result, the surface area of heat islands on the ground and their intensity have increased, particularly in the years 2003, 2006 and from 2009 to 1014. As a result, the minimum temperatures (blue and green) have decreased or even disappeared altogether.

With regard to the distribution of ICU, there are three (03) categories of ICU around the commune of Porto-Novo and its surroundings:
-    Low UHI (yellow colour) with a thermal ëcartage of 1°C with the përiphëries ;
Average UHI (orange colour) with a thermal ëcar! of 2°C with përiphëries ;
-    Strong ICU (red to red colour Бэпсё) with a thermal ëcaЛ of 3°C to 6°C with përiphëries.
In addition, the UHIs of the study area ëyo 1иеп! from the south-west towards the centre and the north-east which corresponds to the area of extension of the city and done of 1 land use. Finally, it should be noted that the spatial variations in tempërature are small and the thermal divide is reduced in our study area. But despite this, there is the presence of UHI whose intensity and spatial variation really depend on the types of land use and the absence or presence of vëgëtation in the commune of Porto-Novo and its surroundings.

In fact, areas of natural vegetation are disappearing at the expense of conurbations and cultivated areas as described in chapter IV. In other words, the dynamics of land use is a function of demographic involution and building density, which are perfectly correlated with the dynamics of UHIs.

Thus, changes in land use (deforestation, urbanisation) in the study area have significant consëquences on local tempëratures, as we can see from the Modis thermal images. As a result, there is a good correlation between the intensity and spatial variation of the UHI and the types of land use.

6.3. Influence of UHI on the population

6.3.1. People's perception of heat

People feel the heat in different ways throughout the day. Figure 68 shows how people feel the heat according to the time of day.

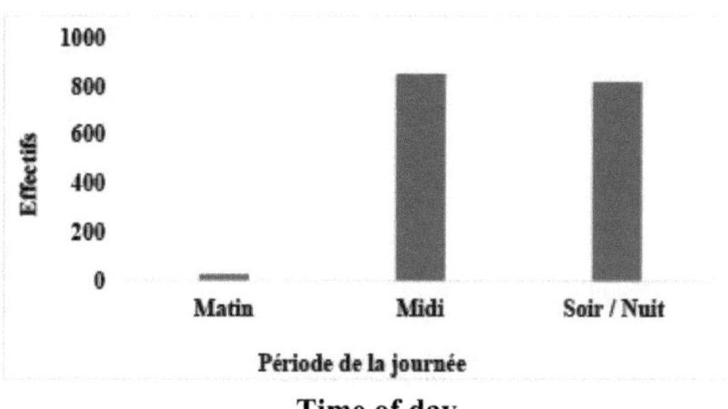

Time of day

Figure 70   Përiode of intense heat during the day **Source:** Data processing, December 2020

Figure 68 shows three periods during which people experience intense heat. Of the 864 people interviewed, 98.80% felt the heat around midday and beyond; 95.20% felt the heat in the evenings around 6pm to 10pm or even 11pm for some, and even midnight for others and beyond; 3.40% felt the heat in the morning from 11am in particular.

It should also be noted that 10 people in this sample (around 1%) said that they felt the heat all day long, and that they were unable to give us an accurate assessment.

The majority of people interviewed, i.e. more than 95%, said that during the day, the heat is felt from around midday until the night. It should be pointed out that all these people acknowledge that the heat is more bearable in the rainy months. They also say that harmattan nights are very cool.

Moreover, according to the surveys, this perception is a function of urban morphology. Wooded streets, green spaces and the sphere of contiguous urban areas modify the microclimate and therefore influence the effects of UHI. Figure 69 shows the proportions of the population who perceive this to be the case, according to location.

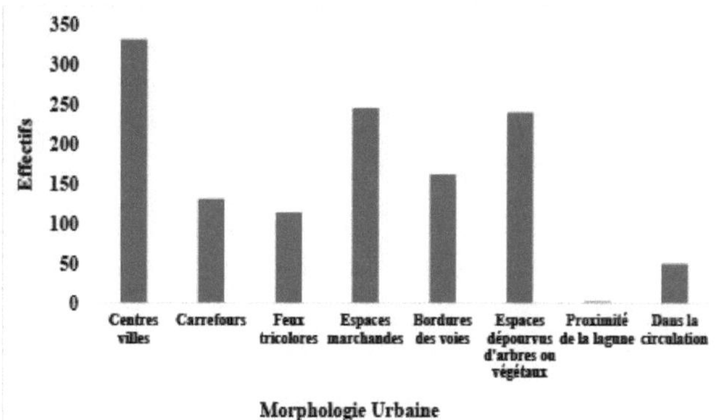

Figure 71   Perception of heat by the population as a function of urban morphology **Source**: Data processing, December 2020

Analysis of Figure 69 shows that built-up and serviced areas are warmer than undeveloped areas. Thus, 48% of respondents said that there are places where it is warmer than others. These include town centres, crossroads, traffic lights, shopping areas, roadsides, tree-lined areas and traffic areas. For more than half of the enquëtës, ьо13ё8 spaces regulate the tempërature and provide coolness. Figure 70 presents the causes of excës of heat.

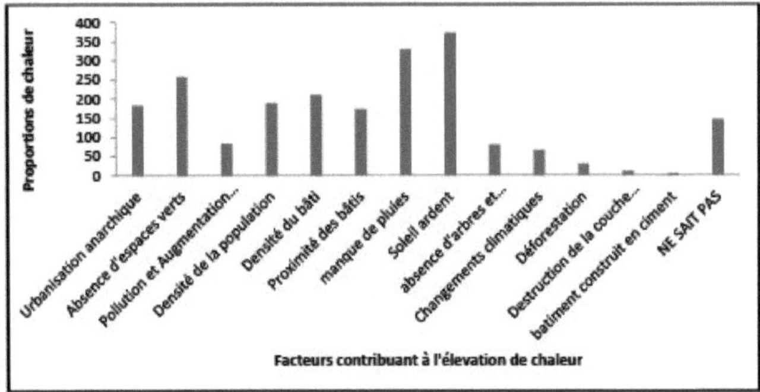

Figure 72: Causes of heat excëses

Source: Data processing, December 2020

Analysis of Figure 70 reveals thirteen (13) causes of heat excëses. The most common causes are: unplanned urbanisation, lack of green spaces, high population density, high building density, lack of or insufficient rainfall and insolation.

6.3.2. Temperature measurement

The tempërature measurement campaign made it possible to observe and assess the variability of the UHI on a conurbation and neighbourhood scale. This campaign also made it possible to relate the population's feelings to a figure and to assess its variability within the same neighbourhood. Table 25 shows Sëries of tempërature measurements. Figures 71 and 72 show respectively the evolution of temparatures for the prerɪ̈ere and second sëries.

Table 25: Sëries of tempërature measurements

1ʳᵉ series of measurements

Sites Jours\^	Site 1 JPN : Jardin des Plantes and Nature	Site 2 Crossroads traffic lights tricolour	Site 3 Lagoon bank	Site 4 Central operation	Timetable 12H-13H30'
For the 5 days	T°max: 26°C T°min : 25 °C	T°max: 33.5°C T°min : 33°C	T°max: 29°C T°min : 29°C	T°max: 32.8°C T°min : 32°C	-

2ᵉᵐᵉ series of measurements

Sites Days	Site 5 In traffic (Porto-Novo city centre)	Site 6 In traffic (Adjarra periphery)	Site 7 Green spaces (town centre)	Site 8 Green spaces (periphery)	Timetable 1.30-3.15 PM
For the 5 days	T°max: 33°C T°min: 32.2°C	T°max: 31 °C T°min: 29.9°C	T°max: 29°C T°min: 28.3°C	T°max: 28°C T°min : 27°C	-

Source: fieldwork, Fëvrier-March 2019

116

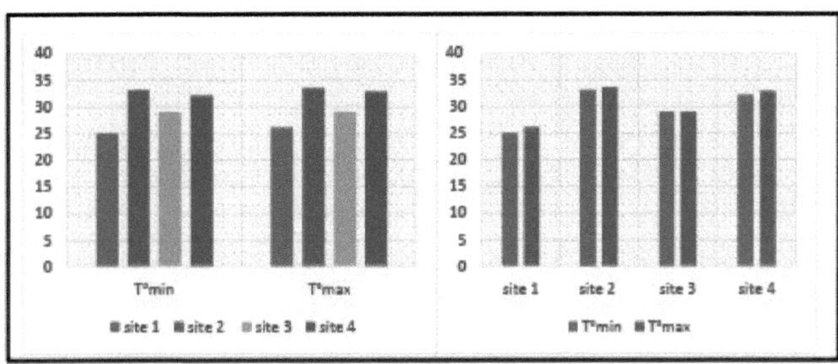

Figure 73: First зёпе of 1етрёгаШгез measurements.

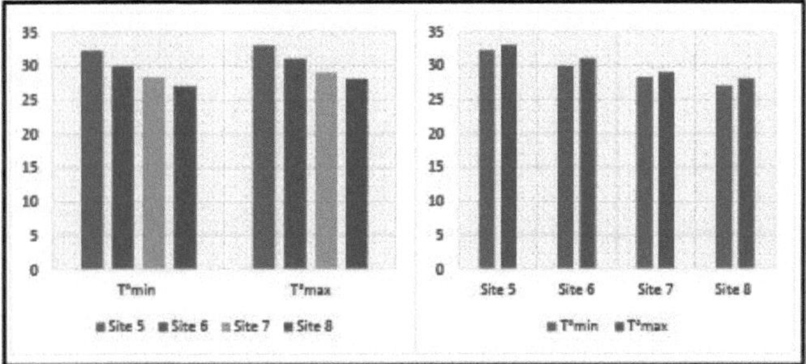

Figure 74: Second series of tempёrature measurements

Looking at Table 25 and Figures 71 and 72, there is significant variability in the intensity of the UHI at each site. The average difference in maximum tempёratures observed during the ргет!ёге sёrie of measurements is around 3°C at the neighbourhood level. In fact, a tempёrature difference of 7.5°C is observed between site 1 (JPN) and site 2 (wando fire), of 3°C between site 1 and and the lagoon bank and finally of 6.8°C between site 1 and the wando initrclie. The same scenario was observed in the second series of measurements. The temperatures observed in the green spaces are significantly lower than those observed in full traffic, with a ёcaЛ of 3 to 4°C. These results confirm people's "comments" that it is warmer in ш^твёв spaces than in vёgёtalisёs spaces. Similarly, these results are also consistent with those obtained using MODIS software. The map below better illustrates this ёstate of affairs (Figure 74).

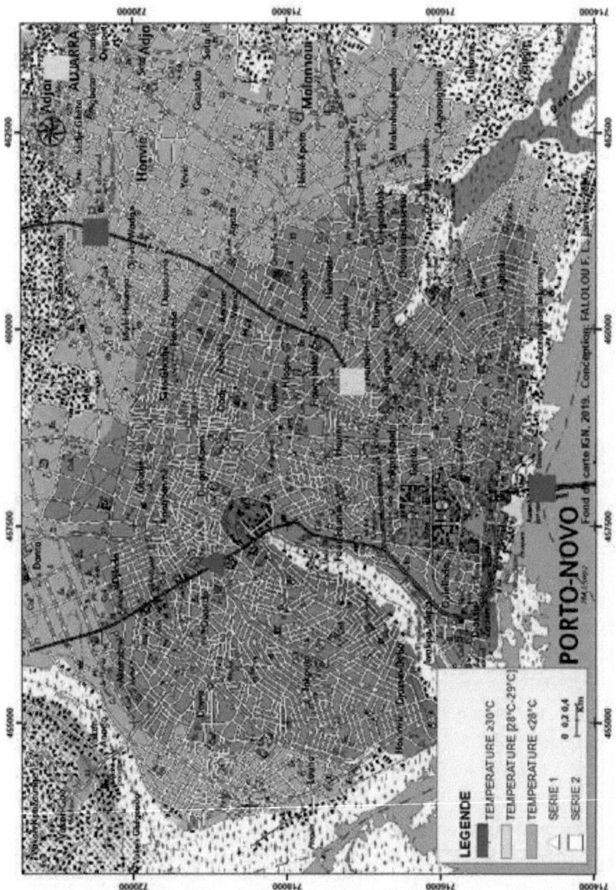

On reading the map, you can see that the 1етрёraШre8 high temperatures (red) are located in areas with уёдё1айоп3, in particular: steps, crossroads with traffic lights, in traffic. The yellow colour indicates medium tempëratures and finally the green colour as for it indicates low tempëratures.

By consëquent, it could be said that the variation and intensity of UHIs dëpend on the presence or absence of green spaces on the one hand and the type of land use on the other.

6.3.2 Impact of heat waves on the population and the environment

"The impact of heat exhausts causes all kinds of discomfort following exposure to the sun"; this is what was stated by the people surveyed, who believe that heat exhausts are increasing every year. They also recognise that this state of affairs has an impact on the socio-economic environment, not to mention the destruction of the ecosystems. The illnesses identified during the surveys include : malaria, skin diseases (measles, chicken pox, pimples, tetcnes on the skin, skin problems in dëpigmentës), fatigue, fever, insomnia, colds, coughs, headaches, insect proliferation, skin irritation, dizziness, skin burns, respiratory difficulties, skin darkening, dehydration and eye irritation. With regard to the economic and environmental impacts,

118

people recognise the following effects following   as dysfunctional: lower yields of food crops; high prices of fruit and some food products; climate change, excës of ройззAёге; stunted growth in vëgëtations.

6.3.3. Proposal for reducing excessive heat by the population.

The solutions recommended by the enquiries relate to socio-anthropological considerations, planning policies and environmental protection. Table XXVI presents these solutions.

Table 26: Solution пропозёез by populations.

Levels	Solutions
Socio-anthropological considerations	- Call in the rainmakers; - Make sacrifices to the *xebiso* divinity *(divinity regulating the rains);* - Sit under the shade of a tree to get some fresh air, - Ban sweeping at night (it chases away the clouds and lengthens the dry season); - Ban menstruating women from going to the swamp.
Development policies	- Promoting reforestation ; - Looking after plants ; - Making building areas viable with an ecological development plan; - More green spaces; - Developing urban forestry ; - Developing public transport
Building architecture	- Spacing out buildings, aerating houses; -
Uses of thermal conditioning equipment	- Use fans and air conditioning;
Use of local materials	- Building with straw and stabilised earth bricks,
Protecting biodiversity	- Banning tree cutting; - Protecting species over a hundred years old; - Prohibit building in sensitive areas (wetlands); - Restore the wetland belt and forests of the lower Oueme delta (in the aguegues, for example) - Protecting and restoring eroded soil
Protecting the environment	- Raising awareness while respecting our environment - Reducing exhaust pollution from motorbikes, vehicles and factories - Reducing greenhouse gas pollution, - Raising awareness of the importance of trees

Source: data processing, December 2020

Table XXIV shows that socio-economic behaviour and the choice of planning policies play an important role in regulating UHI. Non-respect for the environment therefore influences climatic abiances and consequently reduces thermal comfort.

On the other hand, environmentally-friendly development in accordance with architectural standards can enable the ëcosystëmes to perform their functions fully and effectively. The relief and the ëcosystëmes of the environment allow for optimal development: the presence of wetlands, dëpressionary drains and plateaux dotted with basins. Figure 74 shows the major spatial units of the study area through which this layout is perceived.

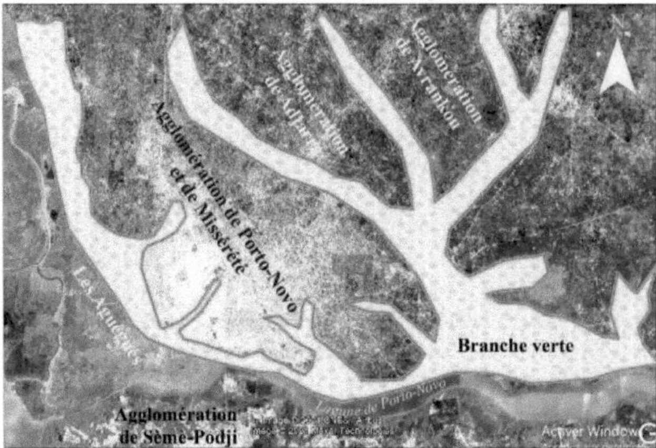

Figure 76: Major spatial groupings in the study area

Source: Google Earth processing & field work

The ëcosytëmiques spaces of the study area can be identified by looking at figure 74. They can be summed up in terms of a dëgradë plateau marked by the agglomerations; the bodies of water are represented by the lagoon, the lake, the marecageous branch of the dëpressions and the lagoon edge. It is this ensemble that needs to be restored and protected in order to guarantee a climate that is safe for the occupants. Nonetheless, green spaces must be built to help regulate the UHI.

6.4. Functions of landscaped areas and plant cover

In the study area, amënagës green spaces are found in urban areas. In Porto-Novo

number of them. number. in fact, these spaces play several

functions. Trees have always been indispensable to man. Immovable and durable, it survives gënërations and bears the imprint of the past. It provides shelter, food, protection, matërials and fuel. All this, and even more, the strong impression made by its size, its timeless yet very much alive side, no doubt explains the special place it occupies, whatever the civilisation, in the human mind.

The green spaces observed in the study area have made it possible to retain the countless functions they provide.

> On the well-being of the individual.

The increase in 1eтpëralllгe in towns compared with the countryside, the high density of ref^ching surfaces on the ground and near buildings, the presence of wind corridors crëёз through high ëdiflces, through streets, the low rate of liumidite проуодиё by 1 insufficiency of plantations and gtizonne surfaces indicate the importance, and even the urgency of introducing vëgëtation into the urban environment by the planting of street trees, by the conservation and amëlioration of existing urban përiurban woodlands (Menviq, 1987). The most

The obvious effect of vëgëtation on the microclimate is shade (photos 9 to 11).

Photo 9: Shade provided by trees at Place Toffa in Porto Novo

Photography: L. FALOLOU, August 2019

Photo 10: Shade created by trees promoting the art and culture exhibition at Place Toffa in Porto-Novo

Photography : L. FALOLOU, December 2021

Photo 11: Shade created by trees at Place Kandevie in Porto-Novo *Photo credit: L. FALOLOU, August 2021*

In these photos, we can see that the trees offer people both shade and places to stroll and relax. Trees absorb and reflect solar radiation so that people seek shade on hot sunny days. The absorption by the vegetation of long-wave radiation from the sun also enables trees to reduce the difference between daytime and night-time temperatures. Under a canopy of trees, days will be cooler and nights cooler. Similarly, the canopy reduces wind speed by offering resistance to air displacement. Wind velocity can be reduced by 50% over a distance of 10 to 20 times the height of the screen formed by the vegetation (Menviq, 1987).

The degree of reduction will depend on the height, thickness and permeability of the vegetation used. Similarly, plant cover also intercepts precipitation and reduces its impact on the soil (reducing splash and erosion).

> **Pair status.**

The presence of wooded areas helps to reduce dust, various chemical pollutants and microbial germs. Dust comes from a variety of sources, both natural and industrial, and can carry chemical pollutants. It comes from traffic and urban activity in general, and carries chemicals and pathogenic (harmful) microbes. The foliage filters the dust to a certain extent, then washes it off the ground when the rain washes it away. The effect of vegetation on the polluted air itself varies greatly from case to case (Menviq, 1987):

> pollutants can be absorbed and transformed by vëgëtation (sulphur dioxide, carbon dioxide and ozone);

> they can be absorbed and accumulated without transformation by the vdgdtal (fluorine, lead). It should also be mentioned that vegetation may have an anti-microbial role. It is now well known that the number of microbial germs per m^2 of air is much lower in a forest than in a city centre street (Menviq, 1987).

> **Water quality.**

The green spaces and vegetation of a city help to absorb rainwater, through percolation (the flow of water into the ground under the effect of gravity) at ground level and through the roots of trees. By preserving green spaces, it is possible to reduce the volume of run-off water, protect water sources and prevent, or at least reduce, the damage caused by flooding. The presence of green spaces also makes it possible to limit the pollution of surface water, which would otherwise flow over paved areas containing pollutants such as lead and waste of all kinds. This dirty water, drained naturally into watercourses or collected by storm drains, contributes to water pollution and the disappearance of aquatic fauna (Menviq, 1987).

> **Soil protection**

Vegetation plays an important role in protecting soil from erosion by water and wind. Left bare, open spaces in urban areas can degrade rapidly. The absence of plant cover makes the soil surface more sensitive to the impact of water droplets (capping effect) and to the force of the wind (Agbossou, 2011).

This can lead to degradation of the soil structure or loss of material (through gullying, erosion, run-off, sludge, sandstorms, etc.). The problem is particularly acute on sloping ground, river banks, cliffs, hills and slopes. It would therefore be advantageous to protect fragile areas, not only by preserving vegetation but also by planting it where it is absent (Menviq, 1987).

> **Noise pollution**

Over the last five decades, average noise levels, especially in cities, have increased considerably. The presence of vegetation can alleviate the discomfort caused by excessive noise levels (Tente, 2008).

Vegetation, with its leaves of varying widths, can reduce noise levels. Studies carried out on the importance of vegetation show that the creation of buffer zones (a strip of wooded land) reduces noise by 6 to 8 decibels per 30 m. This reduction is significant, if we consider that a reduction of 12 decibels corresponds to a reduction in sound sensation of around 50% (Tossou, 2007).

> **A refuge for avian and terrestrial fauna.**

Wooded areas also provide a habitat for a whole range of terrestrial and avian fauna (photos 12 and 13). They play an important role in both natural and periurban environments. Observing wildlife, especially birds, is an increasingly frequent leisure activity. The presence of wooded areas enables this fauna to survive in an urban environment. They therefore play a vital role in the balance of the city's ecosystems. Moreover, the interaction between built-up areas and natural open spaces can also make it possible to develop nature interpretation sites close to urban schools (Menviq, 1987). Photos 12 and 13 show the Jardin des Plantes et de la Nature.

Photo 12: Jardin des Plantes et de la Nature (monkeys peacefully strolling on tree branches)
Photography: L. FALOLOU, December 2019

Photo 13: Jardin des Plantes et de la Nature (monkeys walking around a l'interieur du conservatoire)

Photography: L. FALOLOU, December 2019

> **An architectural and aesthetic element.**

Vegetation also influences the physical expression of the urban environment. It enhances the aesthetics of the built landscape, creating a change in texture and a contrast in colour and form with the adjacent buildings. Around a well-designed building or residence, the vegetation - trees and shrubs - harmonise with and enhance the architectural features. The diversity of foliage and the blossoming of different species add an important note to the built-up areas, which are all too often concentrated and surrounded by vast parking spaces. What's more, in residential areas or public developments, vegetation ensures the private nature of certain species. In addition, retaining a wooded strip can help to isolate a residential area from a major road or an industrial estate. Street planting provides a link between the various public spaces and recreational functions. The city thus becomes a living, well-planned whole. Photos 14 and 15 illustrate this section (Menviq, 1987).

Photo 14: Cite de Grace (Aspect estlietique des Terres Pleins Centraux) in Porto-Novo **Shot**:

Photo 15: TPC Houssou-mëdë in Porto-Novo The harmony of the trees with the town's architecture **Shot**: L. FALOLOU, January 2022

> **Essential social facilities**.

The green spaces serve as recreation areas for relaxation, walking, sport and leisure. Nature interpretation (photos 16 and 17).

Photo 16: JPN, users having fun playing various games - Meeting places and users *Photo taken by: L. FALOLOU, December 2015* d^change des

Their social function stems from the role they play in facilitating access to the public for leisure activities and promoting encounters between citizens.

In short, green spaces play a vital and irreversible role in balancing the ëcosystëmes of a city.

6.5. Anarchic occupation of green spaces and uncivil behaviour

Green spaces are designed for relaxing and unwinding. Therefore, occupying these green spaces for commercial purposes would not be a good idea, according to the people we interviewed. These latterëres pointed out malfunctions detrimental to the sustainability of these spaces. These included:

- risk of dënaturer green spaces ;
- risk turning green spaces into nuirelie ;
- anti-social behaviour and noise pollution: disturbing the peace and quiet of users;
- poor waste management;
- risk dëtourner 1 objectif du lieu ;
- plant stress due to uncivil behaviour ;

In view of these impacts, the local people recommend appropriate measures for the sustainable management of green spaces. They recommend the following measures:
- remunerating hours worked (sweeping and maintenance)
- Empowering users;

In addition, other collective and coercive measures should be considered in order to guarantee the functions of green spaces. Figure 73 shows the distribution of measures according to participation in the management of green spaces.

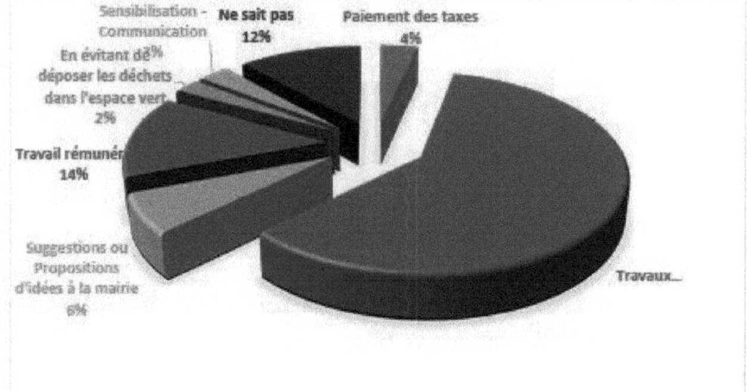

Figure 77: Rëpartition of measures according to involvement in green space management.

Thus, within the populations interviwëes 59.36% have op!ë for community work; 13.74% have op!ë for work гётипёгё, on the other hand 3.51% have prëfërë the final contribution. As far as taxes are concerned, people think this would be an appropriate measure. However, they say they are unaware of the laws governing public management, let alone green spaces.

6.6. Discussion

The present ëtude made it possible to ettra denser des ilots de Chaleur Urbains (ICU) dans la Commune de Porto-Novo et de ses alentours. Analysis of the results reveals the presence of UHIs whose intensity and spatial variation really depend on the types of land use and the absence or presence of vëgëtation in the Commune of Porto-Novo and its surroundings (Lucrece F. *et al.* 2018). Thus, changes in land use (deforestation, urbanisation) in our study area have noticeable consëquences on tempëratures at the local scale as we can see from modis thermal images.

These results confirm those of authors who have analysed the phënomëne of urban heat islands in Beijing, China (Xiao, R. *et al.*, 2007) , Sëoul, South Korea (KIM, Y and BAIK, J., 2005), Baltimore in the United States (LI D. and BOU-ZEID, E. 2013) , Phoenix in the United States (CHOW W., et *al.*, 2012) , Manchester in Great Britain (SKELHORN C. et *al.*, 2014) , Lodz in Poland (Offerle B. et *al.*, 2005) and Montreal in Canada (Bergeron O. and Strachan B. 2012).

Lktude rëalisëe dans la region Parisienne (Madelin M. *et al.*, 2017), montre d'une part que les tempëratures nocturnes de surface des zones densëment urbanisëes sont beaucoup plus ëlevëes (+ 6°C) que les zones agricoles avoisinantes, lors des nuits de ciel clair et d'autre part l'influence notable de 1 occupation/utilisation du sol sur le climat local.

Similarly, these results also confirm those of the study carried out by Rigo G. and Parlow E. (2003) using satellite imagery in the city of Basel in Switzerland. Their results once again highlight the importance of vëgëtation on the thermal behaviour of the study area. Their results concur with those of Ariane S., *et al.* (2012) on Paris, who reaffirm the hypothesis of cooling of the surfaces occupied by urban pares dëja montre par Dousset B. et Gourmelon F., (2003). Thus, the conclusions of these studies corroborate the results obtained for Porto-Novo and the surrounding area in the sense that a link between vëgëtation and tempërature has been established.

Going along these same lines, the study by Cox J. et *al.* (2005) on the тёкорок of New York in the USA, showed on the one hand that at local scales, building densik and street shape played a main role in variations in surface UHI and on the other hand a strong similarity between Involution of vëgëtation and that of temperature. This is in line with Schrijvers P. *et al.* (2015), who claim that the presence of large buildings (urban canyon), the layout of buildings along streets, and the makriaux used for construction and roads are the main causes of nocturnal UHI. Indeed, these makriaux participate in the retention of heat confëranting them a low thermal inertia (CHEN J. *et al.*, 2017). This is the case for asphalt, which is particularly effective at storing heat and gradually releasing it several hours after sunset (Qin Y and Hiller J. 2014). In addition, the rëfërence (Oke T. 1981), attests that the layout of buildings particularly along streets and their height partially impede the movement of fair which participates in the retention of heat in the makriau.

In addition, several ëtudes on urban heat islands using the mëthode of measurement campaign using ктрёгакгез sensors, mëtëorological station, mobile thermometers all lead to the same results. Thus, these diflerentes ëtudes de par le monde dans des villes de failles differentes dëmontrent que la variabilik spatiale de l'ilot de chaleur urbain est fonction des types d'occupation du sol et de 1 absence ou la présence de la yëдёlaйоп, des formes urbaines, des surfaces impermetiNes et de la hauteur moyenne des batis et de leur densite (Xavier F. 2015 ; Efe S. et Eyefia O.2014).

In view of all the above, it should be noted that this increase in tempërature, linked to urbanisation, can have very serious health impacts such as : heat and water stress, discomfort, weakness, impaired consciousness, cramps, syncope, heat stroke, and even exacerbating pre-existing chronic diseases such as diabëte, respiratory insufficiency, cardiovascular, cërëbrovascular, neurological or renal diseases, to the point of causing death (Lachance G., et al., 2006). In this sense, regular monitoring of the phënomëne of the urban heat island and the Involution of land use is fundamental to the sustainable management of the environment and the understanding of its functioning and climatic factors. Thus, as advised by Giguere M. (2009) and the Conseil Rëgional de l'Environnement de Montreal (CRE de Montreal, 2010), increasing the vëgëtal cover of cities by greening and protecting natural spaces is a key factor in reducing urban heat islands.

6.7. A sustainable solutions approach to UHI mitigation

The UHI phenomenon is affecting all the world's cities. Over time, many cities have become aware of the problem of UHI and the issues involved, and they are considering the strategy and actions to be implemented to mitigate this phenomenon.

The results of this study have shown that it is possible to reduce the effects of heat by maintaining as much green space as possible. Thus, to attenuate the said phënomëne, several urban amënagement measures shiver effective. However, the solutions to be developed to deal with urban heat islands must be adapted to the local context and climate.

So, in our study environment :

-     **Awareness-raising campaigns** could be run to raise awareness of the importance of protecting vegetation, whether it's the lawn of one's own garden or the city's parks. In the same way, the importance of choosing light colours for house facades and installing green roofs should also be made clear. Heat islands are the result of urbanisation and must therefore be a priority for the environment and public health.

-     **greening and rainwater management measures;** It is necessary for all new construction to take into account, in its development plan, greening from the first stages of construction. In fact, increasing the vëgëtal coverage of cities through greening and the protection of natural spaces is a key factor in the fight against heat islands. Greening measures have the advantage of being able to be carried out in many areas, such as along roadsides, on public land (municipal land, parks, schoolyards, etc.) and on private land (residential yards, around commercial buildings, parking lots, etc.) (Giguere, 2009; CRE de Montreal, 2010). What's more, greening roofs or walls is an excellent way of countering heat islands. In fact, green roofs, commonly known as green roofs, are good thermal insulators; they reduce the heat inside buildings and homes through evapotranspiration and help to cool the ambient air outside (Giguere, 2009; Labrecque and Vergriete, 2008).

They also help to retain rainwater, reduce noise, increase the longevity of roof membranes, create biodiversity and improve air quality by reducing contaminants (Desjarlais et *al.* 2010). Like green roofs, green walls offer the same benefits. For example, the use of plants adapted to the climate can reduce the temperature of a wall by 20°C compared to a wall with no plants and no shade (Giguere, 2009). Secondly, greening improves rainwater management. In fact, greening the urban environment by planting trees, roofs or green walls increases the city's water retention capacity and increases evapotranspiration, which in turn reduces local temperatures (Boucher, 2010). In addition, increasing the soil permeability rate allows water to infiltrate into the soil and is equivalent to reducing the amount of run-off water that causes natural water to warm up and soil erosion. In addition, if the creation of a planted area is not possible, the use of permeable paving is preferable, because   better infiltration of water into the ground offers a cooling capacity equivalent to that of vegetation (Giguere, 2009).

It should also be noted that the city of Porto-Novo is host to the "Porto-Novo Ville Verte (PNVV)" project, which takes into account urban planning, the development of peripheral areas, the preservation of lagoon bank ecosystems and the city's adaptation to climate change. This project is therefore an asset for the city of Porto-Novo.

Figure 77 below shows a simulation of the ideal green city for our study area.

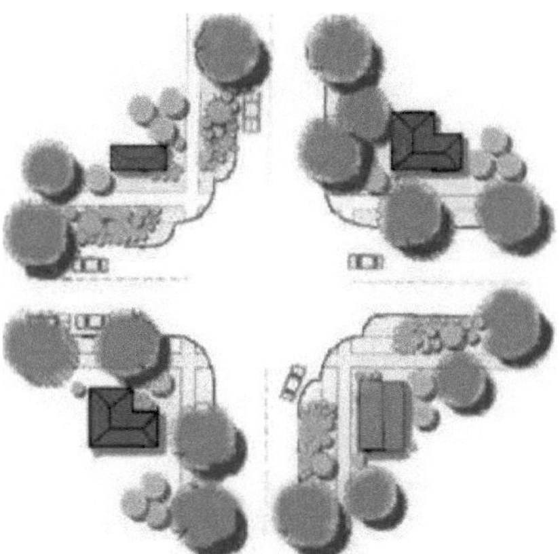

Figure 78: Simulation showing the crossroads of a green city *Source: Google, Geolys, 2022.*

Partial conclusion

The results of the study reveal not only the presence of UHIs in the commune of Porto-Novo and its surroundings, but also that their intensity and spatial variations depend on the types of land use and the absence or presence of vegetation. It should be noted, however, that the spatial variations in tempërature are small and that the thermal difference is reduced in the said study area.

The presence of vegetation and/or watercourses creates cool zones, unlike the hottest surfaces, which are concentrated on building roofs, impermeable surfaces, asphalt and the dark interior of towns.

This ëstudy has enabled us to deepen our knowledge of the thermal behaviour of surfaces in the commune of Porto-Novo and its surroundings during the night in particular.

In order to reduce the effect of these urban heat islands, it would be advisable to implement mitigation measures to counteract the UHI, in order to significantly reduce the fair temperaturëre during the day as well as at night. To this end, there are a variety of actions to reduce this phënomëne such as "vëgëtalisation measures, Hëcз measures to urban infrastructures (architecture and land amënagement), rainwater management and soil permëabilitë actions or measures to reduce anthropogenic heat.

General conclusion

In the city of Porto-Novo and its surroundings, the urban рьёпотеле ёуо1ие from the centre towards the пёпрьёпе. Human activities vary according to each environment. The exercise of these activities often requires precise amënagements and 1 exploitation of natural resources that contribute to the dëgradation of l'environnement. Lktude a rëyëlë that 1 occupation of the ground knew important changes during the different dates: 1972, 1992, 2012. In fact, the ипкёз most widespread land uses are crops and palm groves, crops and fallow land, plantation crops, wetlands, water bodies and agglomerations, which occupied more than 94.72% of the total area in 1972, 95.94% in 1992 and 97.29% in 2012.

In 2032, these classes will occupy more than 96.96% of the total surface area. In general terms, analysis of the various land-use maps has led to the following observations:

- the progression of built-up areas (адд^тёгайо^);
- the decline and even disappearance of sacred forest relics;
- the spread of palm crops and fallow land ;
- the spread of crops and fallow land
- the growth of plantation crops

The classes that appear to be the most stable over these different dates are dënudës soils, water bodies and wetland formations.

Over the various periods 1972-1992, 1992-2012 and 20122032, areas of natural vegetation (remnants of sacred forests) disappeared to the detriment of agricultural areas (crops and jackfruit, palm crops and jackfruit, plantation crops) and built-up areas (agglomerations).

Similarly, it is foundë that agglomerations present a significant positive correlation with dëmography, meaning that the area of agglomerations has increasedë with dëmography. The correlation of dëmography with sacred forests is also significant but negative', meaning that the area of sacred forests has significantly regressed from 1979 to 2012 with 1 population increase in the same përiod.

The other correlation coefficient values are not significant. However, there is a regression in the areas of most of the land-use units with the population growth observed from 1979 to 2012 exceptë for meadows, water bodies and dënudës soils.

Thus, the vëgëtales formations and the cëtural areas are giving way to agglomerations which are constantly expanding due to the evolution of the population.

From all the foregoing, I hypothesis 1 according to which 1 land use undergoes an essentially progressive evolution in the commune of Porto-Novo and its surroundings is confirmedëe. These results allow us to confirm that the параёlrез that could explain the variation in UHIs are notably agglomërations, 1 population increase and 1 land use.

L^tude a ëgalement eопйгтё 1 hypothese N°2 selon laquelle la densite des batis et l'amënagement des espaces verts influencent la variation des ICU. In fact, there is a relationship between the heat quantity measured and the building density observed. Areas of high density should therefore coincide with areas of high heat. Similarly, it has been observed that the presence of green spaces creates a cool zone and thus reduces the temperaturërature of the environment; hence the reduction in the intensity of the UHI.

It can be seen from the above that as the density of buildings increases, so does the intensity of UHI. However, this intensity decreases in the presence of green spaces. Thus, building density and green spaces have a different influence on the variation (intensity) of UHI, one negatively and the other positively.

Similarly, the results of this study revealed not only the presence of UHIs in the commune of Porto-Novo and its surroundings, but also that their intensity and spatial variations depend on the types of land use and the absence or presence of vegetation. It should be noted, however, that spatial variations in tempërature are small and that the heat gap is narrower in the aforementioned research area. The increase in heat varies according to the degree of urbanisation, the type of development, the building (conurbations) and the activities carried out. New activities require people to move, mainly by individual means of transport, which generates atmospheric and noise pollution through engine exhaust fumes. The heat also increases the use of air conditioning in homes, cars and offices. Added to this are the gases generated by the few industries located there. The sanitation probteme is not o^икё with the liquid and solid dëchets which disturb the quiëtude of the populations due to the escaping gases with for consëquence diseases such as: migraines, acute respiratory infections, skin disease, etc.

I .'livpotliese 3 according to which UHI affect the balance of urban climate and populations is conflrmëe. In fact, the presence of vegetation and/or watercourses creates cool zones, unlike the hottest surfaces, which are concentrated on building roofs, impermeable surfaces, asphalt and dark surfaces inside cities. This research has increased our knowledge of the nocturnal and diurnal thermal behaviour of the commune of Porto-Novo and its surroundings. In order to reduce the effect of these urban heat islands, it would be appropriate to implement mitigation measures to significantly reduce the fair temperature during the day and night. To this end, a variety of actions can be taken to reduce this phënomëne, such as "vëgëtalisation measures, urban infrastructure measures (architecture and town and country planning), rainwater management and soil permëabilitë measures, and measures to reduce anthropogenic heat ; This confirms the results according to which the presence of trees (urban green spaces) favours and contributes to the formation of cooler zones in the urban environment which reduce the intensity of the UHI phënomëne.

The basic parameters influencing the urban environment can be summarised as follows:
- the local climate, and more specifically the solar radiation and wind, which are more influenced by the terrain;
- the building environment in its eompiexity, the land and the building materials ;
- anthropogenic heat, produced by human activity: buildings, industry and transport;
- the presence or absence of vegetation directly affects the elements of the climate (tempërature, wind, humidity...) ;

- **Recommendations and outlook**

The main recommendations for combating urban heat islands can be summarised as follows:
- *S* Reinforce planting measures;
J Optimising spatial organisation ;
J Veuillez a l'atteintes des objectifs du projet " Porto-Novo ville verte " afin de capitaliser les résultats aupres des communes satelittes de Porto-Novo. This would enable good practice and applicable results to be managed and exchanged on the basis of the lessons learned from the actions carried out by the aforementioned project.

In the light of the results of this study and the proposed solutions, we intend to extend the study of UHIs in the communes of Abomey-Calavi and Cotonou in order to assist the iintonles in their decision-making, and then to assess the role played by wind circulation in the location of heat islands.

Bibliography

1. **Mourima M.M., (2006).** Pertinence du traitement des images et des SIG dans un тёсашзте d'attente sur les conflits en relation avec la gestion partagёe des ressources naturelles : cas de Bouza et Keita. Мёто!re de DEA gёographie, Fieulte des Lettres et Sciences Humaines, Universite Abdou Moumouni de Niamey, 77p

2. **Ackermann G., Mering C., Quensiere J., (2003).** Analysis of built-up areas extension on the Petite Cote region (Senegal) by remote sensing. *Cybergeo, Revue Europeenne de Geographie*, n° 249, 15 p..

3. **Ademe, (2000).** Classification et critères d'implantation des stations de surveillance de la qualite de l'air. Recommandations du groupe de travail "cttracteristition des sites". Agence de l'environnement et de la maitrise de l^nergie. Ecole des mines de Douai ;

4. **Ademe, (2001).** Le savoir-faire frangais en matiёre de surveillance de la qualite de l'air ambiant, données et rёfёrences. Agence de l'environnement et de la maitrise de l^nergie. ADEME ёditions, Angers ;

5. **Ademe**, **(2006).** Climate change. [On line] *http:/ www.ademe.fr/ changements climatiques ;*

6. **ADG ONG, (2011).** Guide pratique a l'usage des communautes rurales du Delta du Saloum, Sёnёgal

7. **Agence de la sante et des services sociaux de Chaudiere-Appalaches, (2009).** "Whenheat becomes dangerous", Available at at : http://www.santeetenvironnement.ca/. Accessed on 26 June 2009

8. **Ahouandjinou Nathanael O. D., (2004).** "urban pressure on wetlands: the case of the zounvi and boue valleys in porto-novo (benin)", **Memoire de maitrise de geographie,** Universh^ d'Abomey-Calavi (UAC), Бёпт.

9. **Akbari H., Pomerantz M., Taha H., (2001)**, "Cool surfaces and shade trees to reduce energy use and improve air quality in urban areas", *Solar energy*, Vol. 70, pp. 95-310.

10. **Ali Toudert, (2001).** "Mёthodologie d'intёgration de la dimension climatique en urbanisme", in *"Les cahiers de L 'EPAU,"* No. 9/10, Algiers, p. 108.

11. **Aminata D., (2006).** Dynamique de 1 occupation des sols dans des niayes de la region de Dakar de 1954 a 2003 : exemples de la grande niaye de Pikine et de la niaye de Yembeul. Мёто1re de DEA de gёographie, Universite Cheikh Anta Diop de Dakar. (www.memoireonline.com).

12. **Arnfield J.A., (2003).** Two Decades of Urban Climate Research: a Review of Turbulence, Exchanges of Energy and Water, and the Urban Heat Island, *International Journal of Climatology*, 23: 1-26

13. **Assako Assako R.J., (2000).** Mёthode navette image-terrain pour la realisation des cartes d'occupation du sol en milieu urbain africain: le cas de Yaoundё (cameroun). *TELEDETECTION*, vol. 1, p. 285-303.

14. **AtmosphericChemistrydepartment, (2005)** [Online]. *http://www.atmosphere.mpg.de* (Page consulted on 10/05/2014)

15. Barima Y S. S., 2009. Dynamique, fragmentation et diversite vёgёtale des paysages forestiers en milieux de transition foret-savane dans le Dёpartement de Tanda (Cote d'Ivoire). Thёse de doctorat Universite Libre de Bruxelles

16. **Batty M. and Howes D., (2001).** Predicting temporal patterns in urban development from remote imagery. In J. P. Donnay, Barnsley & Longley (Eds.), Remote sensing and urban analysis (pp. 185-204). London: Taylor and Francis.

17. **Behera, M. D., Borate, S. N., Panda, S. N., Behera, P. R. and Roy, P. S. (2012).** Modelling and analyzing the watershed dynamics using Cellular Automata (CA) - Markov model : A geoinformation based approach. Earth System Sciences, 121 (4): 1011-1024.

18. **Besancenot J., (2002).** "Heat waves and mortal^ in large urban agglomërations", *Environnement, risques et sante*, Vol. 4, No. 1, pp. 229-240.

19. **Besancenot, J., (2007).** *"Sante et changement climatique, la montëe de l'inquietude",* Vol. 44, No. 10.

20. Adam S. and Boko M. **(1993).** Le Bënin. Paris, Edicef, 2ëme ëdition, 93 p.

21. **Bessemoulinet P. Olivieri J., (2000)** *Thermal radiation, radiation balance and greenhouse effect,* [Online] *http://www. uwsp.edu/geo/faculty/ritter/glossary/A D/albedo.html* , (Page consulted on 10.01.2006)

22. **Bianchin A., Bravin L., (2004).** Reprodnetibilite des procëdures d'extraction de 1 espace urbain. *Revue Franqaise de Photogrammetrie et de Teledetection*, n° 173/174, p.93-103.

23. **Bianchin A., Bravin L., (2004).** Reproductibilite des procëdures d'extraction de 1 espace urbain. *Revue Franqaise de Photogrammetrie et de Teledetection*, n° 173/174, p.93-103.

24. **Bigot S., Zin I., Diedhiou A., (2005):** Apports de données de HRV de SPOT pour l^tude des variations phdnologiques dans le bassin de l'Оиётё (Бёши). *Teledetection.* Vol 4(4).pp.339-353.

25. **Blanc N., (2007),** *"Vers une esthetique environnementale,* Editions Quae Coll, Nss Indisciplines," Paris. DOI : 10.1051/nss/2009045

26. **Bolund P., Hunhammar S., (1999).** "Ecosystem services in urban areas", *Ecological Economics*, Vol. 29, pp. 293-301.

27. **Bouchard M., and Smargiassi A., (2007).** *"Estimation des impacts sanitaires de la pollution atmospherique au Quebec : essai d'utilisation du air quality benefits assessment tool", (AQBAT),* Institut National de Santë publique du Quebec, 59 p. *Available at this address :* *http://www. inspq.qc.ca/pdf/publications/817 ImpactsSanitairesPollutionAtmos.pdf*

28. **Bourque A., Simonet G., Lemmen, D.S., Warren, F.J., Lacroix, J., Bush, E., (2007).** *Quebec,* Chap. 5. In: "From Impacts to Adaptation: Canada in a Changing Climate", Government of Canada, Ottawa, pp. 171-226.

29. **Brady E., (2007).** "Vers une veritable esthetique de l'environnement: lamination des EonPёreз et des oppositions dans l'expërience esthetique du paysage", *Cosmopolitiques*, 15 pp. 61-72.

30. **Bridier S., Quenol H. and Kermadi S., (2005).** Mëthodes d'analyse de la repartition des tempëratures et de 1 ilot de chaleur urbain a Lyon. *XXth conference of the International Association of Climatology.* Gënes, 85-88.

31. **Bridier S., Quenol H. and Kermadi S., (2005),** "Mëthodes d'analyse de la repartition des tempëratures et de 1 ilot de chaleur urbain a Lyon", *XXeme colloque de l'Association Internationale de Climatologie.* Gënes, 85-88, 2005.

32. **Caloz, R. and Collet, C. (2001)** Precis de Tëlëdëtection, vol. 3 : Traitements numëriques d'images de tëlëdëtection. Presses de I'Universite du Quebec et Agence universitaire de la Francophonie, Sainte-Foy, 386 p.

33. **Cantat O., (2004).** L'ilot de chaleur urbain parisien selon les types de temps. *Norois*, 2, 75-102.

34. **Carrega P, (1994).** "Topoclimatology and habitat. Revue d'Analyse Spatiale Quantitative et Appliqueeu", Thëse d'Etat, vol 35&36, 408p,

35. **Carrega P. (1994).** *Topoclimatology and habitat.* Revue d'Analyse Spatiale Quantitative et Appliqueeu Thëse d'Etat, vol 35&36, 408p,.

36. **Cavayas F. and Baudouin Y., (2008).** *"Etude des biotopes urbains etperiurbains de la CMM, Volets 1 et 2: Evolution des occupations du sol, du couvert végétal et des ilots de chaleur sur le territoire de la Communaute metropolitan de Montreal (19842005)".* Conseil Rëgional de l'Environnement de Laval 123p.

37. **Chen, H. and Pontius, R. G. (2010).** Diagnostic tools to evaluate a spatial land change projection a long a gradient of an explanatory variable. Landscape Ecology, 25 : 13191331.

38. **Chi X., Maosong L., Cheng Z., Shuqing A., Wen Y., Jing M.C., (2007).** "The spatiotemporal dynamics of rapid urban growth in the Nanjing metropolitan region of China" *Landscape Ecology*, n° 22, p.925-937.

39. **Chopin F., Mering C., (2004).** Cartographie de la densite du bati par analyse granulomëtrique des images de tdldddtection. *Revue Franqaise de Photogrammetrie et de Teledetection*, n° 173/174, p.113-122.

40. **Chopin F., Mering C., (2004).** Cartographie de la densitd du bati par analyse granulomdtrique des images de tdldddtection. *Revue Franqaise de Photogrammetrie et de Teledetection*, n° 173/174, p.113-122.

41. **Clarke K.C., Parks B.O. and Crane M.P., (2002).** Geographic information systems and environmental modeling. New Jersey: Prentice Hall.

42. **Colombert M., (2008),** *"Contribution a l'analyse de la prise en compte du climat urbain dans les differents moyens d'intervention sur la ville"*, Universitd Paris-Est, PhD thesis, 538 p.

43. **Economic and Social Council (ESC), (2007).** *Tendances demographiques à echelle mondiale, report by the secretary-general*, 25p. [Online] URL: http://daccess-dds-ny.un.org/doc/UNDOC/GEN/N07/206/12/PDF/N0720612.pdf?OpenElement. Consultd on 3 May 2012.

44. **Conseil Regional de 1 Environnement de Laval, (2008).** (CRE Laval), available at: http://www.cmm.qc.ca/biotopes/ , Consultd on 14 August 2014.

45. **Conseil Regional de 1 Environnement de Montreal** (CRE de Montreal), **(2008).** *"Les materiaux reflechissants et permeables pour contrer les ilots de chaleur urbains"*, 20 p. Available at the following addressaddress: http://www.cremtl.qc.ca/fichiers-cre/files/pdf991.pdf

46. **Conseil Regional de 1 Environnement de Montreal** (CRE de Montreal), **(2010),** *" Guide sur le verdissement pour les proprietaries institutionnels, commerciaux et industriels "*, 42 p. Available at the following address: http://www.cremtl.qc.ca/fichiers-cre/files/SBM2010/Guide Verdissement Entreprises.pdf

47. **Conseil Regional de I'Environnement de Montreal, (2007),** *"Lutte aux ilots de chaleur urbains"*, Le Conseil, Montreal, 54 p.

48. **Coquillard, P. and Hill, D. (1997).** Modëlisation et simulation d'^cosystemes : des modeles dëterministes aux simulations a ëvënements discrets. Masson, 2: 100-124.

49. **Coutts A. M., Beringer J., Tapper N., (2008),** "Changing urban climate and CO2 emissions: implications for the development of policies for sustainable cities", *Urban Policy and Research*, In Press.

50. **CRODT, (2006).** Recensement National de la Peche artisanale maritime sënëgalaise. Rapport final

51. **Deng J.-S., Wank K., Li J., Deng Y.-H., (2009).** Urban Land Use Change Detection Using Multisensor Satellite Images. *PEDOSPHERE*, n° 19 (1), p. 96-103.

52. **Denis Hebert,** April 2014, "(urban heat lots and climate change", Dëpartement de Gëomantique applique ; Facultës des Lettres et des Sciences Humaines ; Universe de Sherbrooke

53. **Ding H., Wang R.C., Wu J. P., Zhou B., Shi Z., Ding L. X., (2007).** "Quantifying Land Use Change in Zhejiang Coastal Region," China Using Multi-Temporal Landsat TM/ETM+ Images. *PEDOSPHERE*, n° 17(6), p. 712-720.

54. DOI : 10.1080/01431169008955053

55. **Donnay J.P. and Unwin D., (2001).** Modelling Geographical Distributions in Urban Areas. In Donnay, Barnsley, Longley (eds.), Remote Sensing and Urban Analysis, pp. 205-224.

56. **Dubreuil V., Montgobert M. et Planchon O., (2002),** " Une mëthode d'interpolation des tempëratures de 1 air en Bretagne : combinaison des paramëtres gëographiques et des mesures infrarouge ", NOAA-AVHRR. *Hommes et Terres du Nord*, no. 1, pp. 2639, 2002.

57. **Dubreuil V., Montgobert M. and Planchon O., (2002).** Une methode d'interpolation des tempëratures de 1 air en Bretagne: combinaison des paramëtres gëographiques et des mesures infrarouge NOAA-AVHRR. *Hommes et Terres du Nord*, no. 1, pp. 26-39.

58. **Durieux L., Lagabrielle E., Nelson A., (2008).** A method for monitoring building construction in urban sprawl areas using object-based analysis of Spot 5 images and existing GIS data. *ISPRS Journal of Photogrammetry and Remote Sensing*, n° 63, p. 399-408.

59. **Durieux L., Lagabrielle E., Nelson A., (2008).** A method for monitoring building construction in urban sprawl areas using object-based analysis of Spot 5 images and existing GIS data. *ISPRS Journal of Photogrammetry and Remote Sensing*, n° 63, p. 399-408.

60. **Eastman, J. R. (2006).** IDRISI Andes. Guide to GIS and Image Processing. Worcester, Clark University, 457 p.

61. **Efe S. I., O. A. Eyefia, (2014).** "Urban Warming in Benin City, Nigeria," Atmospheric and Climate Sciences, 4, 241-252. . http://dx.doi.org/10.4236/acs.2014.42027, Department of Geography and Regional Planning, Delta State University, Abraka, Nigeria

62. **English P., Fitzsimmons K., Hoshiko S., Margolis H., McKone T.E., Rotkin-Ellman M., Solomon G., Trent R., Ross Z., (2007),** "Public health impacts of climate change in California: community vulnerability assessments and adaptation strategies", *report No. 1: heat-related illness and mortality, Information for the Public Health Network in California.* California Department of Public Health, Salt Lake City, 49 p.

63. **Fao, (1995).** Forest resource assessment 1990. Global Synthesis .FAO Rome

64. **Faure J-F., Tran A., Gardel A., Polidori L., (2004).** Elaboration of a population density index and analysis of its spatial distribution in Belem (Brazil) and Cayenne (French Guiana). *Revue Franqaise de Photogrammetrie et de Teledetection*, n° 173/174, p.135-144.

65. **FEDELE C., (2010).** "Adaptation de la ville a 1'augmentation des tempbratures". Etude en droit de 1'urbanisme". Master II Professional thesis "Droit et mbtiers de 1'urbanisme et de 1'immobilier"; UNIVERSITE PAUL CEZANNE AIX- MARSEILLE III FACULTE DE DROIT ET DE SCIENCES POLITIQUES.

66. **Foody G.M., (2002).** "Status of land cover classification accuracy assessment", Remote Sensing of Environment, vol. 80, pp. 185-201. DOI : 10.1016/S0034-4257(01)00295-4

67. **Foody, G.M. (2002).** Status of land cover classification accuracy assessment. Remote Sensing of Environment, 80, pp. 185-201.

68. **Francoual T. (1994).** *Oasis, notice d'utilisation du logiciel.* Paris: Laboratoire de Science des sols et Hydrologie de l'INA-PG, 18 p.

69. **Gar-On Yeh A., Li X., (2001).** Measurement and monitoring of urban sprawl in a rapidly growing region using entropy. *Photogrammetric Engineering and Remote Sensing*, vol. 67, n°1, p.83-90.

70. **Geist, H. J. and Lambin, E. F. (2001).** What Drives Tropical Deforestation? A metaanalysis of proximate and underlying causes of deforestation based on subnational case study evidence. LUCC Report Series, No. 4, 136 p.

71. **IPCC, (2001).** *"Climate Change Assessment: The Scientific Basis"*, Working Report I of the Intergovernmental Panel on Climate Change, Switzerland. [Online] *http://www.ipcc.ch* (Page consulted on 25 March 2004)

72. **Girard C.M., Girard M.C. (1994).** "Aide a la cartographie d'unites paysageres par une mëthode d'analyse du voisinage des pixels: application en Basse Normandie". *Photo-interpretation*, n°3-4, p. 145-154.

73. **Girard, M. C. and Girard, C. M. (1999).** Traitement des données de tëlëdëtection. Dunod, Paris, 529 p.

74. **Givoni B., (1989),** "Urban design in different climates", *WCA-10, WMO/TD,* No. 346, W.M.O.

75. **Haentjens, J., (2008),** "Les villes lievres ; Rendre dësirable le dëveloppement durable, *Futuribles*, n° 342, pp. 49-53.

76. **Hans Wackernagel, (2013),** Basics in Geostatistics 1, Geostatistical structure analysis:The variogram; MINES ParisTech; NERSC, April 2013. http://hans.wackernagel.free.fr

77. **Herold M., Goldstein N.C. and Clarke K.C., (2003).** The spatiotemporal form of urban growth: measurement, analysis and modeling. Remote Sensing of Environment 86(2003)286-302

78. **Herold M., Menz G. and Clarke K.C., (2001).** Remote sensing and urban growth models-demands and perspectives. Symposium on remote sensing of urban areas, Regensburg, Germany, June 2001, Regensburger Geographische Schriften, vol. 35, on supplement CD-ROM.

79. **Herve Quenol, Olivier Vergne and Vincent Dubreuil, (2007).** Anais XIII Simposio Brasileiro de Sensoriamento Remoto, Florianopolis, Brasil, 21-26 avril 2007, INPE, p. 5467-5469. Apport de la dëoтайдие pour la etiraeleristition de l'Ilot de Chaleur Urbain a Rennes (France), Laboratoire COSTEL, UMR6554 LETG University Rennes2, place du recteur Henri Le Moal, 35043 Rennes Cedex (France) herve.quenol@uhb.fr ; vincent.dubreuil@uhb.fr

80. **Hountondji I.C.H., (2008):** Dynamiques environnementales en zones sahëlienne et soudanienne de l'Afrique de l'Ouest : Analyse des modifications et ëvaluation de la dëgradation du couvert vëgëtal. Thëse pour obtenir le grade de Docteur en Sciences de l'univershy de Liëge, Belgique. 153 p.

81. **Hu Z., Lo C. P., (2007)**, "Modeling urban growth in Atlanta using logistic regression" *Computers, Environment and Urban Systems*, n° 31, p. 667-688.

82. **Inglada J., (2001)**. *State of the art in change detection on remote sensing images.* Toulouse: CNES, 20 p.

83. **Jacquin A., Misakova L., Gay M., (2008)**. A hybrid object-based classification approach for mapping urban sprawl in periurban environment. *Landscape and Urban Planning*, n° 84, p. 152-165.

84. **Jean-Marc Jancovici, (2004)**, *"Le rechauffement climatique : reponse a quelques questions elementaires"*, x- Environnement. [Online] *http://www.x-environnement.org,* (Page consultëe le 4 November 2004)

85. **Jensen J.R. and Cowen D.C., (1999)**. Remote sensing of urban/suburban infrastructure and socioeconomic attributes. Photogrammetric Engineering and Remote Sensing, 65, 5:611 - 622.

86. **Katrina Ao, Hanh Ngo T.M, (2006)**. *"A GIS Analysis of Vancouver's Urban Heat Island. "*, Part resultsand section, *www.geog.ubc.ca/courses/klink/g470/class00/kf0/pointb.htm* (Page consultëe on 16 May 2013).

87. **Kressler F., Kim Y., Steinnocher K**. **(2003)**. "Object-oriented land cover classification of panchromatic KOMPSAT-1 and SPOT-5 data". *Proceedings of IGARSS 2003 IEEE*, Toulouse.

88. **Lachal B., (1995)**. *"Quelques aspects du climat urbain de Geneve et ses consequences sur l'environnement"*, p. 119, in Lachal B, Romerio F, Weber W, (Ed.), Actes de la journee du CUEPE N° 62, Universite de Geneve'.

89. **Lacroix V., Idrissa M., Hincq A., (2005)**. Spot 5 for urbanisation detection. *Revue Frangaise de Photogrammetrie et de Teledetection*, n° 178, p.3-11.

90. Lambin, E.F. et al., 2001. The causes of land-use and land-cover change: moving beyond the myths. Global Environmental Change, 11, pp. 261-269.

91. **Leconte F, petrissans M. (2014)**. Caratitiisation des ilots de chaleur urbaines par zonage climatique et mesures mobiles : cas de Nancy. p.274. Doctoral school Ressources Procëdës Produits Environnement. Universite de Lorraine.

92. **Leroux Louise, (2012)**. "Analyse diachronique de la dynamique paysagère sur le bassin supërieur de l'Oиëtë (Benin) a partir de l'imagerie Landsat et MODIS- Cas d^tude du communal de Djougou", Rapport 2012, Hydrosciences Montpellier, ANR ESCAPE.

93. **Loireau M. (1998)** : Espaces - Ressources - Usages : Spatialisation des interactions dynamiques entre les systëmes sociaux et les systëmes ëcologiques au Sahel nigërien. Thëse Universite Montpellier III - Paul Valery 411 p.

94. **Lopez-Paniagua, J., F. J. Romero y A. Velazquez. (1996)**. Human activities and their impact on the habitat of the zacatuche conejo. In: Velazquez, A., F.J. Romero y J. Lopez-Paniagua (Eds). Ecologia y conservation del conejo zacatuche Romerolagus diazi y su habitat. Fondo de Cultura Economica/ Universidad Autonoma de Mëxкo, pp. 119-132.

95. **M.G.O.P. Obasi, (2001)**. *"Cities accentuate climate revolution*, WMO", [Online] *http://www.wmo.ch/web/Press/citiesfr.html* (Page eonsriltee le 15.02.2006)

96. **Maestripieri, N. (2012)**. Spatiotemporal dynamics of industrial ГоreзЁëreз plantations in southern Chile. De l'analyse diachronique a la modëlisation prospective. Thëse de doctorat, Universite Toulouse 2 Le Mirail, Toulouse, France. 357 P.

97. **Maestripieri, N. and Paegelow, M. (2013)**. Spatial validation of two simulation modëles: l'example of industrial plantations in Chile. Cybergeo: European Journal of Geography, Systëmes, Modëlisation, Gëostatistics. 653, 30 p.

98. **Mahamane A. (2007)**: Analyse diachronique de 1 occupation des terres et caratieristiques de la vëgëtation dans la commune de Gabi (region de Maradi, Niger) pp.296-304.

99. **Mahe G. Olivry J.C. Desouassi R. Orange D. Bamba F. and Servat E. (2000):** Relation Eaux de surface-eaux souterraines d'une riviere tropicale au Mali, C.R. Acad. Sci., Sciences de la Terre et des Pkinetes, 330, 689-692.

100. **Major D. J., Baret F., Guyot G., (1990).** "A ratio vegetation index adjusted for soil brightness", International Journal of Remote Sensing, vol. n° 5, p. 727-740.

101. **MAS J.F., (2000)** "A review of tdldddtection methods and techniques of change". *Canadian Journal of Remote Sensing*, vol. 26, no. 4, pp. 349-362.

102. **Mas, J.-F., Kolb, M., Houet, T., Paegelow, M. and Camacho-Olmedo, M. T. (2011).** Informing the choice of tools for simulating changes in land use and occupancy patterns. Geomatics and land use, pp. 405-430.

103. **Masek J.G., Lindsay F.E., Goward S.N., (2000)**, "Dynamics of urban growth in the Washington DC metropolitan area", 1973-1996, from Landsat observations. *International Journal of Remote Sensing*, vol. 21, no. 18, p. 3473-3486.

104. **Maurice G. Estes, Jr, Virginia Gorsevski, Camille Russell, Dale Quattrochi, Jeffrey Luvall (2014).** "The Urban Heat Island Phenomenon and Potential Mitigation Strategies and Potential Mitigation Strategies," Alabama, p 4. *http://www.ghcc.msfc.nasa.gov* (Page accessed January 15, 2014).

105. **McKee, J.K.; Sciulli, P.W.; Fooce, C.D.; Waite, T.A. (2003).** Forecasting global biodiversity threats associated with human population growth. Biological Conservation, 115, pp. 161-164.

106. **Melissa Giguere, M. Env, (2009)**, "Biological risk management, environmental and occupational "; Institut national de sante publique du Quebec

107. **MEPN-DPN, (2011).** Plan d'amenagement et gestion de 1 aire marine protégee de Cayar 2011 -2015. January 2011

108. **Michard R., (1966).** Le Soleil, Paris, p. 69, 71, 72.

109. **Moise Tsayem Demaze and Alain Trebouet, (2008).** "Mapping and evaluation multi-echelle de l'etalement urbain a l'aide d'images Spot XS : Exemple du Mans (Ouest-France) ", UMR CNRS ESO-GREGUM, Universite du Maine, Avenue O. Messiaen 72085 Le Mans cedex 09.

110. **Morgane Colombert and Philippe Boudes, (2012)**, "Adaptation aux changements climatiques en milieu urbain et approche globale des trames vertes ", *VertigO - la revue electronique en sciences de l'environnement* [En ligne], Hors-serie 12 | mai 2012, publié en ligne le 15 mai 2012, consulte le 11 aout 2014. URL: http://vertigo.revues.org/ 11821; DOI: 0.4000/vertigo.11821

111. **Morgane Colombert, Jean-Luc Salagnac, Denis Morand and Youssef Diab,** (2012) . "Le climat et la ville : la пёесввкё d'une recherche croisant les disciplines", *VertigO - la revue electronique en sciences de l'environnement* [En ligne], I lors-serie 12 | mai 2012, uploaded 15 May 2012, accessed 12 Aug 2014. URL : http://vertigo.revues.org/11811 ; DOI : 10.4000/vertigo.11811

112. **Moulinie C., Naudin-Adam M., (DUAT), (2005).** "quick note on l'occupation du sol" №383, institut d'amenagement et d'urbanisme de la region d'Ile d e

France.

113. **MOUSSA M. S., (2006).** Systëme d'information (SIG) et dynamique de l'occupation du sol du bassin versant du kori Goubë degre carre du Niger. Mëтolre de DEA gëographie, Ftierilte des Lettres et Sciences Humaines, Universite Abdou Moumouni de Niamey, 73p.

114. **N'bessa B.,** (1997). "Porto-Novo et Cotonou (Bëпт): origine et ëyolиPoп d'un urban doublet". Bordeaux III. Universite Michel de Montaigne : Thëse de Doctorat d'Etat es Lettres.416 pages and appendices.

115. **Nikolopoulou, M., (2004).** *Designing outdoor spaces for the environment a bioclimatic approach.* Center for Renewable Energy Sources, 64 p.

116. **Noyola-Medrano, M.C. (2006).** Current morphological evolution of the Champ Volcanique de la Sierra Chichinautzin (Mexique) a partir de l'analyse tomomorphomëtrique des cones de scories et du changem ent de l'occupation du sol. Tesis Doctorat. Universite Paris 7 Denis-Diderot.

117. **Oke T.R., (1978).** *Boundary layer Climates.* Methuen & Co, London, 372 p.

118. **Oke T.R., (1988).** "Street design and urban canopy layer climate", *Energy and Buildings,* Vol. 11, pp. 103-113.

119. **Oke, T.R. (1981).** "Canyon geometry and the nocturnal urban heat island: comparison of scale model and field observations". *International Journal of Climatology*, nol, p. 237- 254.

120. **Oke, T.R., (1987)** "*Boundary layer climates.*" 2nd ed., Routeledge, London, 474 p.

121. **Oke, T.R., (1988)** "The urban energy balance Progress" in *physical geography*, Vol. 12, pp 471-508.

122. **Oloukoi J, Vincent JM. ;t AGBO, FB. (2006)** modëlisation de la dynamique de 1 occupation des terres dans le dëpartement des collines au bëтп in journal canadien de tëlëdëtection vol. 6, n° 4, p. 305-323

123. **Oszwald J., Lefebvre A., Arnault DE Sartre X., Thales M., Gond V., (2010)** : Analysis of the directions of change in ëtats de surface yëдë!aиx to inform the dynamics of the magaranduba pioneer front (para, bresil) between 1997 and 2006. *Teledetection journal.* Vol 9(2). pp. 97-111.

124. **Ouattara T., Dubois J.M., Gwyn J., (2006).** "Mëthods for the mapping of. l'occupation des terres en milieu aride a l'aide de donnèes multi-sources et de l'indice de vëgëtation TSAVI', Tëlëdëtection, vol. 6, no. 4, pp. 291-304.

125. **OUSSEINI I., (1994).** " Rëpartition spatiale de 1 occupation humaine et ressources naturelles dans la region du fleuve Niger, in Au contact Sahara-Sahel Milieux et sociëtës du Niger, Revue de gëographie alpine, vol 2, pp 159-170.

126. **Tools   Solar, (2009).** *Solar Glossary* : http://www.outilssolaires.com/Glossaire/default.htm. Consultë on 3 April 2009.

127. **Paegelow, M. (2004).** Gëomatics and gëography of the environment: From l'analyse spatiale a la modëlisation prospective. Habilitation a Diriger des Recherches, Universite de Toulouse - Le Mirail, France. 211 p.

128. **Paegelow, M. and Camacho-Olmedo, M. T. (2005).** Possibilities and limits of prospective GIS land cover modelling - a compared case study : Garrotxes (France) and Alta Alpujarra Granadina (Spain). International Journal of Geographical Information Science, 19 (6) : 697-722.

129. **Paegelow, M., Villa, N., Cornez, L., Ferraty, F., Ferre, L. and Sarda, P. (2004).**

Modёlisations prospectives de 1 occupation du sol. The case of a mёditerranёenne mountain. Cybergeo : European Journal of Geography, Systёmes, Modёlisation, Gёostatistiques. n°295, 20 p. [online] http://cybergeo.revues.org/2811 (half visit : 03/12/2013).

130. **Park M.H., Stenstrom M.K., (2008).** Classifying environmentally significant urban land uses with satellite imagery. *Journal of Environmental Management*, n°86, p.181-192. Phёnomёne d'ilots de chaleur en milieu urbain montreal, 4 mai 2010 ecole de technologie зирёпеиге university du диёbec тётоие prёзеп!ё a l'ёcole de technologie superieure

131. **Philippe Anquez and Alicia Herlem, (April 2011).** "Heat islands in the Montreal metropolitan region: causes, impacts and solutions", Universite du Quebec A Montreal (UQAM)

132. **Pickup G., and Marks A., (2000)** "Identifying large-scale erosion and deposition processes from air-borne gamma radiometrics and digital elevation models in a weathered landscape", Earh Surface Processes and Landforms, vol. 25, p. 535-557.

133. **Pigeon, G., Lemonsu A., Masson V., Hidalgo, J., (2008).** "De l'observation du microclimat urbain a la modёlisation intёgrёe de la ville", *La Meteorologie*, No. 52, pp. 39-47.

134. **Pontius, R. G. (2000).** Quantification error versus location in comparison of categorical maps. Photogrammetric Engineering and Remote Sensing, 66 (8): 10111016.

135. **Pontius, R. G. (2010).** Workshop Land Change Modeling Methods: calibration, validation and extrapolation. SAGEO'10 - Spatial Analysis and Geomatics. Toulouse. [online] http://hal-emse.ccsd.cnrs.fr/docs/00/35/63/42/PDF/OPDE08_MBH_FP.pdf (first visit : 13/12/2013)

136. **USAID/COMFISH PENCOO GEJ Project, (2012).** Gestion concertёe pour une peche durable au Sёnёgal, " *dynamique de 1 occupation des terres, cartographie des CLPA, des zones de peches et mise en place d'un systeme d'information geographique (SIG)* " Exёcution report, July 2012.

137. **Puissant A., Weber C., (2004).** Dёmarche orientёe "objets-attributs" et THRS image classification. *Revue Franqaise de Photogrammetrie et de Teledetection*, n° 173/174, p.123-134.

138. **Rahim Aguejdad, (2009).** Urban sprawl and ё assessment of its impact on the biodiversite, de la reconstitution des trajectories _a la modёlisation prospective. Application _a une agglomёration de taille moyenne : Rennes Metropole. Geography. Univershy Rennes 2, 2009. French. <tel-00553665>

139. REPAO, Analyse des pratiques, des politiques et des institutions de peche et les climate change in Sёnёgal

140. **Resources   Ressources naturellesCanada,   2004**, "A   time of change :   the changements climatiques au Quebec, un climat en constante transformation", Accessible at http://adaptation.nrcan.gc.ca/posters/qc/qc 02 f.php Оопвикё on 13 November 2008.

141. **Saaty, T. L. (1990).** How to make a decision: The Analytic Hierarchy Process. European Journal of Operational Research, 48: 9-26.

142. **Salomon T., Aubert C.,** 2004, "*La fraicheur sans clim*", Terre Vivante, Paris, 160 p.

143. **SANDA GONDA Hassane, (2010).** "Cartographie de la dynamique de l'occupation des sols et de 1'erosion dans la ville de Niamey et sa pёriphёrie", Mёmoire de Maitrise en gёographie 2010, Universite Abdou Moumouni de Niamey, Facultё des Lettres et Sciences

Humaines / Dëpartement de Gëographie.

144. **Sarr M.A., (2009).** Cartographie des changements de 1 occupation du sol entre 1990 and 2002 in northern Sënëgal (Ferlo) from Landsat images. *Cybergeo : European Journal of Geography* [Online], Environment, Nature, Pay sage, article 472, online 07 October 2009, accessed 30 October 2012. URL : http://cybergeo.revues.org/22707 ; DOI : 10.4000/cybergeo.22707

145. **SARR M.A., 2009**: Cartographie des changements de 1 occupation du sol entre 1990 and 2002 in northern Sënëgal (Ferlo) from Landsat images. Cybergeo : European Journal of Geography [Online], Environnement, Nature, Paysage, article 472, online 07 October 2009, accessed 30 October 2012. URL : http://cybergeo.revues.org/22707 ; DOI : 10.4000/cybergeo.22707

146. **Sebastien Bridier, Anais XIII Simposio Brasileiro de Sensoriamento Remoto, Florianopolis,** Brasil, 21-26 avril 2007, INPE, p. 5467-5469. Apport de la gëomatique pour la cara^risation de 1'Ilot de Chaleur Urbain a Rennes (France), UMR ESPACE, Universite de Provence, Avenue Robert Schuman, 13 Aix en Provence (France) Sebastien.bridier@up.univ-mrs.fr

147. **Sietchiping R., 2003.** Evolution of the urban space of Yaoundë, Cameroon, between 1973 and 1988 by tëlëdëtection. *TELEDETECTION*, vol. 3, n° 2-3-4, p. 229-236.

148. **Skupinski G., Binh Tran D., Weber C.,** 2009. Spot multi dates and spatial mëtrics in the study of urban and suburban change - The case of the lower Bruche valley (Bas-Rhin, France). *Cybergeo, Revue Europeenne de Geographie,* n° 439, 22 p.

149. **Soares-Filho, B. S., Pennachin, C. L. and Cerqueira, G. (2002).** DINAMICA - a stochastic cellular automata model designed to simulate the landscape dynamics in an Amazonian colonization frontier. Ecological Modelling, 154 (3) : 217-235.

150. **Taibou Ba and Dieynaba Seek. 2012.** Dynamique de L'Ocupation des sols , cartographie des CLPA, des zones de peche et mise en place d'un systeme d'information geographique. Centre de Suivi Ecologique and USAID/COMFISH Project, Senegal, University of Rhode Island, Narragansett RI, 66 pp.

151. **Tanina Drissa Soro, Bernard Dje Kouakou, Ernest Ahoussi Kouassi, Gbombele Soro, Amani Michel Kouassi, Konan Emmanuel Kouadio, Marie-Solange Oga Yei and Nagnin Soro, 2013.** "Hydroclimatologie et dynamique de 1'occupation du sol du bassin versant du Haut Bandama a Tortiya (Nord de la Cote d'Ivoire)", Les ëditions en environnement : VERTIGO, Volume 13 Nuinero 3 / dëcembre 2013

152. **Thompson, R. S.; Anderson, K. H.; Bartlein, P. J. 1999.** Atlas of Relations Between Climatic Parameters and Distributions of Important Trees and Shrubs in North America. U.S. Geological Survey, Professional Paper 1650, part A and part B.

153. **Trottier A., (2008).** "*Toitures vegetales: implantation de toits verts en milieu institutionnel"*, *Etude de cas :* Universe du Quebec A Montreal (*UQAM),* 80 p.

154. **UICN, (2007)** Mangrove du Sënëgal, Charte de gestion, Rapport final : les mangroves du Sënëgal ; situation actuelles des ressources, leur exploitation, leur conservation .

155. **United States Environmental Protection Agency (USEPA), (2008).** a, " Ground-level ozone: health and environment" USEPA. Available at: http://www.epa.gov/air/ozonepollution/health.html . Consuh^ June 26, 2014.

156. **United States Environmental Protection Agency** (USEPA), **(2008).** b, " *Reducing urban heat islands: compendium of strategies, urban heat island basics"*, USEPA, Washington, DC, 19 p.

157. **Vagen, T.G. (2006)**. Remote sensing of complex land use change trajectories-a case study from the highlands of Madagascar.　　　Agriculture, Ecosystems & Environment, 115(1-4), pp. 219-228.

158. **Velazquez A., et al, (2002)**. "Patrones y tasas de cambio de uso del suelo en Мёхко". Gaceta Ecologica, 62, pp. 21-37.

159. **Verburg, P. H., Soepboer, W., Veldkamp, A., Limpiada, R., Espaldon, V. et Sharifah Mastura, S. A. (2002).** Modeling the Spatial Dynamics of Regional Land Use: the CLUE-S Model. Environmental Management, 30 (3): 391-405.

160. **Verheij R. A**, (1995), "Explaining urban-rural variations in health: a review of interactions between individual and environment " Social Science and Medicine 42, pp. 923-935

161. **Vincent Dubreuil, Herve Quenol, Vincent Nedelec, Jean Francois Mallet, Laurent Durieux and Gilda Maitelli, (2008), "** ëtude de l'impact du changement de l'occupation du sol sur les tempëratures dans la region d'Alta Floresta, Bresil " ; Bulletin de la Sociëtë gëographique de Liëge, 51, 2008, 79-90

162. **Vitousek, P.M., Mooney, H.A., Lubchenco, J., Melillo, J.M. (1997)**. Human domination of Earth's ecosystems. Science, 277, pp. 494-499.

163. **Voogt J. A., (2002).** "Urban heat island, *Encyclopedia of global environmental change*" Vol. 3, pp. 660-666.

164. **White F., (1983).** The vegetation of Africa. UNESCO, Paris. 356 p.

165. **Xavier FOISSARD, 2015,** "The urban heat island and change climatique : application a l'agglomëration rennaise", Thëse de doctorat a I'Universite de Rennes 2 haute-Bretagne. 247p.

166.

167. **Xian G., Crane M., McMahon C., 2008**. Quantifying multi-temporal urban development characteristics in Las Vegas from Landsat and ASTER data. *Photogrammetric Engineering and Remote Sensing*, vol. 74, n°4, p.473-481.

168. **Xu H., 2007**. Extraction of urban built-up land features from Landsat imagery using a thematic-oriented index combination technique. *Photogrammetric Engineering and Remote Sensing*, vol. 73, n°12, p.1381-1391.

169. **Schlaepfer　　R. (2002)** :　　　Analysis of　　　dynamiquedu paysage.　　Laboratoire de gestion des dcosystdmes (GECOS), Ecole polytechnique iederale de Lausanne, Lausanne, 11 p.

170. **Schlaepfer R. (2003)**: Cours Ecologie du Paysage : une introduction, 53 p.

171. **Djafarou ABDOULAYE, 2014** "Dynamique De L'Occupation Des Terres Et Ses Incidences Sur L'ecoulement Dans Le Bassin De L'oueme A L'exutoire De Beterou (Nord-Benin)", Thëse thesis, Uieulle des Lettres, Arts et Sciences Humaines (FLASH) / Universite d'Abomey-Calavi (U.A.C.)

172. **Kouassi Jean-Luc, 2014 "**Suivi De La Dynamique De L'Occupation Du Sol A L'aide De L'imagerie Satellitaire Et Des Systemes D'informations Geographiques : Cas De La Direction Regionale Des Eaux Et Forets De Yamoussoukro (Cote D'Ivoire)" mdmoire de Thëse / Institut National Polytechnique / Ecole Supërieure d'Agronomie (ESA).

173. **Z. Wan, Yulin Zhang, Qincheng Zhang, Zhao-liang Li , 2002, "**Validation of the land-surface temperature products retrieved from Terra Moderate Resolution Imaging Spectroradiometer data", Remote Sensing of Environment, Nov 2002, 83:163180

174. *Z. Wan, Y. Zhang, Z. Li, R. Wanga, V. V. Salomonsonb, A. Yvesc, R. Bossenoc and J. F. Hanocqd*, **2002** "Preliminary Estimate of Calibration of the Moderate Resolution Imaging Spectroradiometer Thermal Infrared Data Using Lake Titicaca," Remote Sensing Environment, Vol. 80, No. 1, 2002, pp. 497-515. doi:10.1016/S0034-4257(01)00327-3

175. **E. T. Crosman and J. D. Horel, 2009** "MODIS-Derived Surface Temperature of the Great Salt Lake," Remote Sensing of Environment, Vol. 113, No. 1, 2009, pp. 7381. doi:10.1016/j.rse.2008.08.013 [23]

176. **C. Coll, Z. Wan and J. M. Galvem, 2009** "Temperature-Based and RadianceBased Validations of the V5 MODIS Land Surface Temperature Product," Journal of Geophysical Research, Vol. 114, No. D20, 2009, Article ID: D20102.

177. **Z. Wan and Z.-L. Li, 2008,** "Radiance-based Validation of the V5 MODIS Land-surface Temperature Product," Internal Journal of Remote Sensing, Vol. 29, No. 17-18, 2008, pp. 5373-5395. doi: 10.1080/01431160802036565

178. **Chu, D. A., Y. J. Kaufman, C. Ichoku, L. A. Remer, D. Tanre, and B. N. Holben, (2002),** Validation of MODIS aerosol optical depth retrieval over land, Geophys. Res Lett, 29(12), doi:10.1029/2001GL013205.

179. **Petrenko, M. and Ichoku, C. (2013),** Coherent uncertainty analysis of aerosol measurements from multiple satellite sensors, Atmos. Chem. Phys. 13, 6777-6805, doi: 10.5194/acp-13-6777-2013.

180. **Reed B.C., Brown J-F., Vander Zee D., Loveland T.R., Merchant J.W., Ohlen D.O. (1994),** Measuring phonological variability from satellite imagery, 703- 714pp.

181. *Velazquez et al.2002*

182. **AMADOU M. SANNI et al, (2009):** Villes du sud : Dynamiques, diversites et enjeux dëmographiques et sociaux. Ed des archives contemporaines. 369 p.

183. **TRIBILLON, J. F., (1992):** Instruments d'amdnagement et de gestion fonciere urbaine africaine. 262 p premiere version.

184. Revue de gdographie du laboratoire Lei'di - ISSN0051 - 2515 -N°11, December 2013 , P- 267-276p Pression ddmographique et degradation de l'environnement dans le departement du Couffo au Benin FANGNON Bernardi BABADJIDE Charles Lambert2 GONZALLO Germain 3 et TOHOZIN Antoine Yves4 / 1

185. (DYNAMIQUE DE L'OCCUPATION DU SOL ET EVOLUTION DES TERRES AGRICOLES DANS LA COMMUNE DE SINENDE AU NORD-BENIN , Gildas Louis Djohy , Henri Sourou Totin Vodounon 1, 2 Nickson Esther Kinzo 1 Ddtails , 1 Departement de Gdographie et Amdnagement du Territoire, 2 LACEEDE - Laboratoire Pierre Pagney, Climat, Eau, Ecosysteme et Ddveloppement, Type de document : Article dans une revue, Cahiers du CBRST, Centre Bdninois de la Recherche Scientifique et Technique 2016, pp.101- 121, Domaine : Sciences de l'environnement / Environnement et Soeiete.

186. International Journal of Biological and Chemical Sciences, Impacts des activitds humaines sur les ressources forestieres dans les terroirs villageois des communes de Glazone et de Dassa-Zoume au centre-Bdnin, ; B Tente, MA Baglo, JC Dossoumou, H Yedomonhan, Vol 5, No 5 (2011) >

187. Merlin, Pierre and Frangoise Choay eds. 1988. Dictionnaire de 1 urbanisme et de l'amdnagement. Paris: Presse Universitaires de France.

188. **Rouxel F., 1999:** "I'hdritage urbain et la ville de demain. Pour une approche de ddveloppement durable. Ministere de I'dquipement, des transports et du logement. Direction Generale de l'urbanisme, de l'habitat et de la construction.

189. Johanna B., Catherine M. and Corinne V., 2014. "Can we map

taches urbaines a partir d'images Google Earth?", Cybergeo : European Journal of Geography [En ligne], Dossiers, document 682, online 24 July 2014, accessed 19 July 2018. URL : http://journals.openedition.org/cybergeo/

190. The bandwidth swept by the satellite in one pass is 2300 km http://terra.nasa.gov/About/MODIS/modis swath.php

191. Direction regionale de 1'Environnement, de 1'Amenagement et du Logement de Bretagne, July 2013, "La densitd et ses perceptions : Modalites de calcul de la densite", RAPPORTS, Service: Climat Energie Amenagement Logement, Division: Amenagement, Urbanisme et Logement, Unite: Amenagement et urbanisme durable.

192. Agence d'urbanisme de Marseille, July 2009, "Density and urban form". http://www.agam.org/fileadmin/ressources/agam.org/publications/manifestations/Dens it%C3%A9 et formes urbaine 2009.pdf Work by 1 agence de Marseille, on how urban forms can make density more pleasant, on how to densify existing spaces. Examples of operations: "densifier et aerer, densifier et être chez soi en ville, densifier et eco construire" Typologie d'habitat selon les COS. No calculation elements.

193. Centre for Studies on Networks, Transport, Urban Planning and Construction publiques (CERTU), July 2002, "Densite: Concept, exemples et mesures", 9 rue Juliette Recamier 69-456 Lyon cedex 06.

Webography

1. World Conservation Union (IUCN): WWW.IUCN.org/brao

2. Direction regionale de 1'Environnement, de l'Amenagement et du Logement Bretagne: www.bretagne.developpement-durable.gouv.fr

3. Department of   demography at the University of Montreal (Canada): www.demoumontreal.ca

4. Institut de Recherche pour le Developpement (IRD): www.ird.fr

5. United Nations Population Fund (UNFPA): www.unfpa.org

6. www.google.fr

7. URL farm : http://www.mystartsearch.com/?type=sc&ts=1416941922&from=amt&uid=ST500LT012-9WS142_W0V4QQBKXXXXW0V4QQBK

8. www.afs-journal.org

9. www.tregouet.org

10. http://www.iaurif.org

11. www.techno-science.net , consulted on 29/10/2015

12. (www.linternaute.com consulted on 22/10/2015).

13. (http://www.guide-clea.fr/clea projet/facteur-de-vue-du-ciel/ , consulted on 20/10/2015 at 15H45.)

14. (www.actu-environnement.com consulted on 21/10/2015).

15. . (www.notre-planete.info consulted on 22/10/2015).

16. www.aquaportail.com consulted on 21/10/2015)

17. www.larousse.fr consulted on 21/10/2015).

18. www.envstudies.brown.edu/classes/es201/2003/forestry/

19. http://gen%C3%A8ve/expose-espace-vert-methodologie.htm. Accessed on 21/10/2012 at 13 hours 42 minutes.

20. http://m.futura-sciences.com/magazines/ , consulted on 21/10/2015

21. (www.actu-environnement.com consulted on 21/10/2015).

Appendices

Annexes 1

Correlations: Density, Agglomeration

Pearson correlation of Densite and Лд1отёгайоп = 0.993

P-Value = 0.007

Correlations: Density, Cultures_jacheres

Pearson correlation of Densite and Cи1Шге8_]асHёге8 = 0.543

P-Value = 0.457

Correlations: Density, Cultures_jacheres a palmier

Pearson correlation of Densite and Culturesjticlieres a palm tree = -0.881

P-Value = 0.119

Correlations: Density, sacred_forests

Pearson correlation of Densite and forets_sacrees = -0.999

P-Value = 0.001

Correlations: Density, marecages

Pearson correlation of Densite and marecages = 0.656

P-Value = 0.344

Correlations: Density, Culture a plantation

Pearson correlation of Densite and Culture a plantation = -0.855

P-Value = 0.145

Correlations: Density, Water level

Pearson correlation of Densite and Plan d'eau = 0.638

P-Value = 0.362

Correlations: Density, Plantation

Pearson correlation of Densite and Plantation = -0.715

P-Value = 0.285

Correlations: Densite, Sol_denude

Pearson correlation of Densite and Sol_dënudë = 0.523

P-Value = 0.477

Appendix 2: Average height and circumference of trees

plots	height (m)	Conference	diameter	lnDl) Y=exp (-1.996 + 2.32	LnDl	
1	12,75	80	25,477707	248,565128	3,23780383	5,5157049
1	12	145	46,1783439	987,745809	3,83251094	6,89542539
1	12	117	37,2611465	600,430341	3,61795113	6,39764663
1	12,55	34	10,8280255	34,143066	2,38213772	3,53055952
1	12,25	87	27,7070064	301,964645	3,32168532	5,71030994
1	10,5	65	20,7006369	153,54314	3,03016447	5,03398157
1	10,5	46	14,6496815	68,8445555	2,6844186	4,23185114
1	10,55	49	15,6050955	79,7124828	2,7475975	4,3784262
1	12,75	107	34,0764331	488,024691	3,52860603	6,190366
1	9	162	51,5923567	1277,45692	3,94337354	7,1526266
1	6	37	11,7834395	41,5431506	2,46669511	3,72673266
1	6,75	20	6,36942675	9,96921634	1,85150947	2,29950198
1	6,15	55	17,5159236	104,210834	2,86311039	4,64641609
1	6,75	17	5,41401274	6,83774543	1,68899054	1,92245806
1	6,09	17	5,41401274	6,83774543	1,68899054	1,92245806
1	3,75	53	16,8789809	95,6294038	2,82606911	4,56048034
1	5,25	103	32,8025478	446,738922	3,49050619	6,10197436
2	12	89	28,343949	318,314371	3,34441357	5,76303948
2	15,75	170	54,1401274	1428,60751	3,99157564	7,26445548
2	14,25	70	22,2929936	182,347109	3,10427244	5,20591207
2	15,75	170	54,1401274	1428,60751	3,99157564	7,26445548
2	18,75	160	50,955414	1241,16577	3,93095102	7,12380636
2	17,25	160	50,955414	1241,16577	3,93095102	7,12380636
2	15	66	21,0191083	159,079186	3,04543194	5,06940211
3	24,75	875	278,66242	63933,7404	5,63000109	11,0656025
3	8,25	72	22,9299363	194,662736	3,13244332	5,2712685
3	15	135	42,9936306	836,846441	3,76105198	6,72964059
3	15	77	24,522293	227,47303	3,19958262	5,42703168
3	12,75	49	15,6050955	79,7124828	2,7475975	4,3784262
3	21,75	700	222,929936	38097,711	5,40685754	10,5479095
4	14,25	107	34,0764331	488,024691	3,52860603	6,190366
4	16,5	139	44,2675159	895,500365	3,79025113	6,79738263
4	14,25	125	39,8089172	700,007448	3,68409094	6,55109097
4	4,5	110	35,0318471	520,358269	3,55625757	6,25451755
4	16,5	76	24,2038217	220,677962	3,18651054	5,39670445
4	8,25	75	23,8853503	213,999896	3,17326531	5,36597553
4	15	136	43,3121019	851,298166	3,76843209	6,74676244
4	9	57	18,1528662	113,21425	2,89882847	4,72928205
4	15,75	190	60,5095541	1849,18254	4,10280127	7,52249895
5	6,75	86	27,388535	293,973266	3,3101245	5,68348883
5	6,75	97	30,8917197	388,670746	3,43048818	5,96273257
5	21	212	67,5159236	2384,35341	4,21236347	7,77668326
5	6	30	9,55414013	25,5383445	2,25697458	3,24018103
5	8,25	48	15,2866242	75,9890651	2,72697821	4,33058945
5	12	100	31,8471338	417,130005	3,46094739	6,03339794
5	12	100	31,8471338	417,130005	3,46094739	6,03339794
5	21	265	84,3949045	4001,30685	4,43550703	8,2943763
5	21	280	89,1719745	4546,50909	4,4905668	8,42211498

5		17,25	76	24,2038217	220,677962	3,18651054	5,39670445
5		12	66	21,0191083	159,079186	3,04543194	5,06940211
6		24,75	382	121,656051	9346,70278	4,80119781	9,14277892
6		14,25	85	27,0700637	286,10361	3,29842846	5,65635402
6		17,25	160	50,955414	1241,16577	3,93095102	7,12380636
6		18	220	70,0636943	2598,31649	4,24940475	7,86261901
6		18	138	43,9490446	880,624798	3,78303089	6,78063165
6		7,5	32	10,1910828	29,6632943	2,3215131	3,3899104
6		7,5	38	12,1019108	44,194617	2,49336336	3,78860299
6		11,25	72	22,9299363	194,662736	3,13244332	5,2712685
6		9,75	65	20,7006369	153,54314	3,03016447	5,03398157
6		10,5	70	22,2929936	182,347109	3,10427244	5,20591207
6		11,25	86	27,388535	293,973266	3,3101245	5,68348883
6		15	75	23,8853503	213,999896	3,17326531	5,36597553
6		22,5	350	111,464968	7629,73218	4,71371035	8,93980802
6		10,5	40	12,7388535	49,7795092	2,54465665	3,90760344
7		14,25	95	30,2547771	370,331108	3,40965409	5,91439749
7		13,5	105	33,4394904	467,122304	3,50973755	6,14659112
7		14	96	30,5732484	379,437885	3,42012539	5,93869091
7		13,25	100	31,8471338	417,130005	3,46094739	6,03339794
7		14,1	95	30,2547771	370,331108	3,40965409	5,91439749
7		14,25	108	34,3949045	498,671494	3,53790843	6,21194755
7		14,14	93	29,6178344	352,494115	3,38837669	5,86503393
7		13,45	110	35,0318471	520,358269	3,55625757	6,25451755
7		13,5	103	32,8025478	446,738922	3,49050619	6,10197436
7		15,75	105	33,4394904	467,122304	3,50973755	6,14659112
7		12	32	10,1910828	29,6632943	2,3215131	3,3899104
7		12,75	48	15,2866242	75,9890651	2,72697821	4,33058945
7		12,75	56	17,8343949	108,659488	2,88112889	4,68821903
7		20,25	115	36,6242038	576,886559	3,60070933	6,35764564
7		9	57	18,1528662	113,21425	2,89882847	4,72928205
7		9	62	19,7452229	137,600544	2,98291159	4,92435488
7		6,75	59	18,7898089	122,644506	2,93331464	4,80928997
8		21	243	77,388535	3272,48742	4,34883864	8,09330565
8		18	86	27,388535	293,973266	3,3101245	5,68348883
8		19,5	105	33,4394904	467,122304	3,50973755	6,14659112
8		20,25	187	59,5541401	1782,14878	4,08688582	7,4855751
8		20,25	88	28,0254777	310,078197	3,33311401	5,73682451
8		7,5	28	8,91719745	21,7609606	2,18798171	3,08011757
8	6,75		32	10,1910828	29,6632943	2,3215131	3,3899104
8	8,25		67	21,3375796	164,72707	3,06046982	5,10428998
8	5,25		42	13,3757962	55,745496	2,59344682	4,02079662
8	13,5		46	14,6496815	68,8445555	2,6844186	4,23185114
8	21,75		102	32,4840764	436,740864	3,48075001	6,07934003
8	21,75		87	27,7070064	301,964645	3,32168532	5,71030994
8	22,5		315	100,318471	5975,19269	4,60834984	8,69537163
8	22,5		87	27,7070064	301,964645	3,32168532	5,71030994
8	21		325	103,503185	6424,52211	4,63960238	8,76787753
9	24		468	149,044586	14970,6897	5,0042455	9,61384955
9	24		392	124,840764	9924,19082	4,82703904	9,20273057
9	24		376	119,745223	9009,6357	4,78536634	9,10604992
9	24,05		406	129,299363	10765,9369	4,86213036	9,28414243
9	23,75		325	103,503185	6424,52211	4,63960238	8,76787753
9	21		125	39,8089172	700,007448	3,68409094	6,55109097
9	21,15		188	59,8726115	1804,33694	4,09221916	7,49794846

9	21	250	79,6178344	3495,36273	4,37723812	8,15919243
9	17,25	86	27,388535	293,973266	3,3101245	5,68348883
9	18,75	92	29,2993631	343,763047	3,37756578	5,8399526
9	21	105	33,4394904	467,122304	3,50973755	6,14659112
9	17,25	86	27,388535	293,973266	3,3101245	5,68348883
9	22,5	445	141,719745	13318,8478	4,95385148	9,49693544
9	21,75	225	71,656051	2737,37863	4,2718776	7,91475604
10	21,75	415	132,165605	11327,7333	4,88405572	9,33500927
10	14,25	97	30,8917197	388,670746	3,43048818	5,96273257
10	16,5	225	71,656051	2737,37863	4,2718776	7,91475604
10	15,75	175	55,7324841	1527,98716	4,02056317	7,33170656
10	14,25	125	39,8089172	700,007448	3,68409094	6,55109097
10	15	106	33,7579618	477,508424	3,51921629	6,1685818
10	15	82	26,1146497	263,220414	3,26249645	5,57299176
10	12,75	55	17,5159236	104,210834	2,86311039	4,64641609
10	20,25	175	55,7324841	1527,98716	4,02056317	7,33170656
10	17,25	100	31,8471338	417,130005	3,46094739	6,03339794
10	16,5	87	27,7070064	301,964645	3,32168532	5,71030994
10	23,25	315	100,318471	5975,19269	4,60834984	8,69537163
10	22,5	430	136,942675	12300,372	4,91956241	9,41738479
10	19,5	96	30,5732484	379,437885	3,42012539	5,93869091
11	12,75	96	30,5732484	379,437885	3,42012539	5,93869091
11	17,25	45	14,3312102	65,4220973	2,66243969	4,18086008
11	6,75	20	6,36942675	9,96921634	1,85150947	2,29950198
11	17,25	124	39,4904459	687,083849	3,67605877	6,53245634
11	19,5	110	35,0318471	520,358269	3,55625757	6,25451755
11	19,55	118	37,5796178	612,403533	3,62646182	6,41739143
11	21	310	98,7261146	5757,45511	4,5923495	8,65825083
11	14,25	105	33,4394904	467,122304	3,50973755	6,14659112
11	19,5	208	66,2420382	2281,27882	4,19331528	7,73249145
11	11,25	62	19,7452229	137,600544	2,98291159	4,92435488
11	15,75	87	27,7070064	301,964645	3,32168532	5,71030994
11	9	65	20,7006369	153,54314	3,03016447	5,03398157
11	9,05	70	22,2929936	182,347109	3,10427244	5,20591207
	2029,08	18926				

I want morebooks!

Buy your books fast and straightforward online - at one of world's fastest growing online book stores! Environmentally sound due to Print-on-Demand technologies.

Buy your books online at
www.morebooks.shop

Kaufen Sie Ihre Bücher schnell und unkompliziert online – auf einer der am schnellsten wachsenden Buchhandelsplattformen weltweit! Dank Print-On-Demand umwelt- und ressourcenschonend produziert.

Bücher schneller online kaufen
www.morebooks.shop

Printed by Books on Demand GmbH, Norderstedt / Germany